Niranjanamurthy M, Sheng-Lung Peng, Naresh E, Jayasimha S R,
Valentina Emilia Balas (Eds.)
Advances in Industry 4.0

De Gruyter Series on Smart Computing Applications

——

Edited by
Prasenjit Chatterjee, Dilbagh Panchal, Dragan Pamucar,
Sarfaraz Hashemkhani Zolfani

Volume 5

Advances in Industry 4.0

Concepts and Applications

Edited by
Niranjanamurthy M, Sheng-Lung Peng, Naresh E,
Jayasimha S R, Valentina Emilia Balas

DE GRUYTER

Editors

Dr. Niranjanamurthy M
M S Ramaiah Institute of Technology
Department of Computer Applications
Bangalore 560054
India
niruhsd@gmail.com

Dr. Sheng-Lung Peng
National Taipei University of Business
Department of Creative Technologies and
Product Design
Taipei City 10051
Taiwan
slpeng@ntub.edu.tw

Dr. Naresh E
M S Ramaiah Institute of Technology
Department of Information Science and
Engineering
Bangalore 560054
India
nareshkumar.e@gmail.com

Dr. Jayasimha S R
Visvesvaraya Technological University
RV College of Engineering
Department of Computer Applications
Bangalore 560059
India
jayasimhasr@rvce.edu.in

Dr. Valentina Emilia Balas
Aurel Vlaicu University of Arad
B-dul Revolutiei 77
310130 Arad
Romania
balas@drbalas.ro
valentina.balas@uav.ro

ISBN 978-3-11-072536-0
e-ISBN (PDF) 978-3-11-072549-0
e-ISBN (EPUB) 978-3-11-072553-7
ISSN 2700-6239

Library of Congress Control Number: 2022930630

Bibliographic information published by the Deutsche Nationalbibliothek
The Deutsche Nationalbibliothek lists this publication in the Deutsche Nationalbibliografie;
detailed bibliographic data are available on the Internet at http://dnb.dnb.de.

© 2022 Walter de Gruyter GmbH, Berlin/Boston
Cover image: monsitj / iStock / Getty Images Plus
Typesetting: Integra Software Services Pvt. Ltd.
Printing and binding: CPI books GmbH, Leck

www.degruyter.com

Preface

Industry 4.0 is a set of technological changes to create a coherent framework to be introduced in the manufacturing process. A simplistic definition of Industry 4.0 is the "application of the IoT, cloud computing, cyber-physical systems (CPS), and cognitive computing into the manufacturing and service environment. Automation and connectivity within the manufacturing world is not new. Physical to digital (taking physical actions and converting that to digital records) and digital to digital (sharing insights using AI) have also been a part of manufacturing for years now.

The first revolution moved from manual to mechanical by harnessing steam and water power, replacing raw muscle.

The second revolution took advantage of electricity and the assembly line to generate mass production.

The third revolution took off using electronics and software, the most iconic example of these two coming together being the computer – and the Programmable Logic Controller (PLC) in the industrial setting. This gave rise to an era of high-level automation.

The fourth revolution is the revolution happening right now, according to this study. The physical world, the digital world, and the virtual world are colliding together, creating smart products, procedures and processes, and smart factories. The study goes on to describe this revolution being brought on primarily by cyber-physical systems and the Internet of Things to create a highly intelligent, integrated, and automated manufacturing ecosystem that spans far beyond the traditional "factory floor."

This book covers all the four revolution concepts. This book has 14 exhaustive chapters on the concept and applications of Industry 4.0 and relevance as follows:

Chapter 1 discusses the Artificial Intelligence and Machine Learning for Industry 4.0, Industry 4.0 is transforming conventional manufacturing into smart manufacturing and generating new possibilities, where machines learn to understand these systems, connect with the world and adapt their behavior intelligently.

Chapter 2 discusses Industry 4.0: A road map to custom Register Transfer Level design. Time to market with an agile strategic approach for the development and production of products has led mankind to deliberately think or work in more anticipated way. One of the most recurring ways of production in today's industry is through automating almost whole development procedures.

Chapter 3 Describes the Industry 4.0: Driving the digital trans-formation in Banking Sector.Industry 4.0 or Digital Revolution is changing the way we live, changing associations with customers and or-generations, which unavoidably infer that both existing busi-ness techniques and monetary administrations are not excluded from this change.

Chapter 4 describes the Blockchain for IoT and Big Data Applications: A Comprehensive Survey on Security Issues. In this chapter, two such modern technologies

https://doi.org/10.1515/9783110725490-202

that have now become an inevitable part of the human life are discussed along with what security measures they lack, why or how they lack these security measures and what can be done to secure these. But before going to their security aspects, let's first know something about the technologies like what role do these play in society and what is their importance in daily human lives.

Chapter 5 Describes the BECON – A Blockchain and Edge Combined Network for Industry 4.0. With the quantity of internet-connected devices increasing and as IoT (Internet of Things) is gaining popularity, the degree of data and information collected by IoT sensors is incredibly high and needs large amounts of resources for processing and analysing the data. Edge processing permits the sensor data to be processed nearer to its source. This chapter performs a survey on various implementations and researches made by professionals and institutions to analyse this innovative alternative. Also propose the utilization of a blockchain-based localized application to modify IoT edge processing.

Chapter 6 represents the Internet of Things enabled health monitoring system for smart cities. Prior to Internet of Things (IoT), interactions of patients with doctors or physicians had been restrained to visits, textual and tele communications. There was no chance doctors or physicians could monitor patients' health constantly and give advice accordingly. IoT-empowered devices have made remote screening in the healthcare sector promising; unleash the ability to keep patients protected and healthy and engaging doctors to convey the standout care.

Chapter 7 Aims at explain the Smart Farming solution using Internet of Things for Rural area. With growing populace throughout the world, farming and production of food progressively profitable and prepared to do exceptional returns in constrained time. The scope for guide experimentation, viability evaluation thru trial and blunders and many others are now not feasible

Chapter 8 Illustrate the Assistance Device for Visually Impaired Based on Image Detection and Classification using DCNN. Based on the statistics World Health Organization (WHO) has provided, there are around eighty five million visually disabled people worldwide. Among them thirty-nine million people are completely blind. The major difficulties faced by them are obstacle detection and this can be eradicated using Deep Learning. Deep Learning provides computer vision to the system which makes decisions based on training algorithms.

Chapter 9 describes novel feature Manufacturing and investigation of patially glazed geopolymer tile. The present invention provides a process for the preparation of partially glazed Geopolymer tiles by using fly1ash, ground granu-lated blast1furnace slag (GGBS), alkaline1solution and super plas-ticizers. The trial mixes are prepared by1varying the mix proportion of GGBS1and fly ash added with alkaline solution of specified mo-larity, sodium1hydroxide and1sodium silicate were1used to prepare the alkaline solutions for the mixture. The strength of the hardened composites is tested to choose the best mix of required strength for the preparation of tiles.

Chapter 10 describes the application of Water Sensitive Urban Design. Water Sensitive Urban Design is emerging as a crucial way of conserving water as a resource in India. Water-sensitive urban design is a land planning and engineering design approach which integrates the urban water cycle, including storm water, groundwater and wastewater management and water supply, into urban design to minimize environmental degradation and improve aesthetic and recreational appeal. A study in detail of Chennai Water Crisis and solutions for it, and comparing the water management in Mysore discussed.

Chapter 11 Discussed the Degradation of Groundwater Resources By Waste Exposed Land use In Urban Cities. Urbanization has led to the generation of large quantity of domestic solid waste. A part of it is properly managed and the rest is dumped illegally around societal community varies the natural resources of groundwater and the soil which has hazardous impact on biotic environment. Analysis of the samples are carried out for assessing the impact of solid waste on groundwater characteristics and check there permissible limits are within the following guidelines of WHO and BIS, if not proper segregation of solid waste need to be done at the source, minimizing the generation of waste and improving the methods of handling it. In this chapter discussed the degradation of Groundwater Resources By Waste Exposed Land use In Urban Cities.

Chapter 12 describes the Use of Effective Placing Different Lateral Load Resisting Structural System in High Rise RC Frame For All Seismic Zones By Using Response Spectrum Method.

Chapter 13 describes the Implementation of a Semi-automatic Approach to CAN Protocol Testing for Industry 4.0 Applications.

Chapter 14 describes Smart cradle system for Industry 4.0. A cradle system is going to help the parents to rest, even if both of them go to work or woman is housewife, taking care of a baby is stressful. When people are in stress it creates a bad environment around the baby, so being stress free is most essential.

Dr. Niranjanamurthy M, M S Ramaiah Institute of Technology, INDIA

Dr. Sheng-Lung Peng, National Taipei University of Business, TAIWAN

Dr. Naresh E, M S Ramaiah Institute of Technology, INDIA

Dr. Jayasimha S R, RV College of Engineering, Bangalore, INDIA

Dr. Valentina Emilia Balas, Aurel Vlaicu University of Arad, ROMANIA

Contents

Preface —— V

Pramod Sunagar, Darshana A Naik, Shruthi G
Artificial Intelligence and Machine Learning for Industry 4.0 —— 1

Siba Kumar Panda, Konasagar Achyut, Swati K. Kulkarni
Industry 4.0: A Road Map to Custom Register Transfer Level Design —— 21

Ram Singh, Rohit Bansal, Vinay Pal Singh
Industry 4.0: Driving the Digital Transformation in Banking Sector —— 51

Sahana. P. Shankar, Deepak Varadam, Harshit Agrawal, Dr. Naresh E
**Blockchain for IoT and Big Data Applications: A Comprehensive Survey
on Security Issues** —— 65

Rithvik Vasishta S, Tanmay Shishodia, Utkarsha Verma, D Prasanna Sai Rohit,
Dr. J Geetha
BECON – A Blockchain and Edge Combined Network for Industry 4.0 —— 87

Abhishek K L, M Niranjanamurthy, Yogish H K
Internet of Things Enabled Health Monitoring System for Smart Cities —— 107

Yogish H K, M Niranjanamurthy, Abhishek K L
Smart Farming Solution using Internet of Things for Rural Area —— 119

Vaishnavi A S, Dr. Sumana
**Assistance Device for Visually Impaired based on Image Detection
and Classification using DCNN** —— 131

Chethan R, Meghana D S, S Kumar
Manufacturing and Investigation of Partially Glazed Geopolymer Tile —— 155

Jyothi M. R, Dr. H. U. Raghavendra, Rahul Prasad, Priyanka B. N
Water Sensitive Urban Design —— 169

Jyothi. M. R, Dr. H U Raghavendra, Dr. Siddegowda
**Degradation of Groundwater Resources By Waste Exposed Land Use In Urban
Cities** —— 179

Santhosh D, Nambiyanna B, Harish M L, DR R Prabhakara
Use of Effective Placing Different Lateral Load Resisting Structural System in High Rise RC Frame for All Seismic Zones By Using Response Spectrum Method —— 189

Sandeep Kakde, Pavitha U S, Veena G N, Vinod H C
Implementation of A Semi-Automatic Approach to CAN Protocol Testing for Industry 4.0 Applications —— 203

Dwarakanath G V, Dr P Ganesh, Sinchana S R, Ashwini C R
Smart Cradle System for Industry 4.0 —— 215

Editors Details —— 233

Index —— 235

Pramod Sunagar, Darshana A Naik, Shruthi G

Artificial Intelligence and Machine Learning for Industry 4.0

Abstract: Industry 4.0 is transforming conventional manufacturing into smart manufacturing and generating new possibilities, where machines learn to understand these systems, connect with the world and adapt their behavior intelligently. Artificial intelligence, powered by a unique interaction mode between man and machine, has revolutionized industry activity patterns. Intelligent factories consume automated mechanisms and provide digital enablers that allow machinery, via an IoT framework, to communicate to one another and the factory systems on the whole. AI and ML applications in Industry 4.0 have expanded beyond our estimates. Smart manufacturing is characterized as entirely integrated, cooperative manufacturing structures that react in real-time to increasing demands and instances in the intelligent factory, the supply network and customer expectations. Autonomous vehicles and the development of robots are still in the development and testing phases, but ML is used to learn the autonomous vehicle world. The new software products are in great demand for ML and AI.

Keywords: artificial intelligence, machine learning, deep learning, industry 4.0, internet of things, autonomous vehicles, smart manufacturing, smart factory

1.1 What is Artificial Intelligence?

The demand for Artificial Intelligence (AI) is increasing, and we communicate and interact with AI technologies, consciously or unknowingly. In the post-humanist view, human and non-human interactions are pervasive, and it can be debated that there is a symbiotic, intertwining relationship between human and non-human entities. Furthermore, it can argue that the lines between human and non-human beings

Acknowledgment: This work was supported by M S Ramaiah Institute of Technology, Bangalore-560054, and Visvesvaraya Technological University, Jnana Sangama, Belagavi-590018.

Pramod Sunagar, Dept, of CSE, M S Ramaiah Institute of Technology (Affiliated to VTU), Bangalore, India, e-mail: pramods@msrit.edu
Darshana A Naik, Dept, of CSE, M S Ramaiah Institute of Technology (Affiliated to VTU), Bangalore, India
Shruthi G, Dept, of ISE, M S Ramaiah Institute of Technology (Affiliated to VTU), Bangalore, India

https://doi.org/10.1515/9783110725490-001

are blurring, and we are looking forward to a world where technical uniqueness is unavoidable. It can say that singularization has begun to occur at a naive level through expert systems in a wide range of fields. Education is one such area undergoing a digital transformation, and artificial intelligence and derivative technological opportunities have begun to be taken advantage of AI. The existing state of the art and the widespread use of AI apps require us to rethink how AI is used in education to respond to what we are doing and where we are going. Although an extensive and thorough notion, intelligence is also a famously elusive one. More than 70 distinct concepts are identified and checked in their detailed survey of available meanings of intelligence. They describe intelligence by removing the most common characteristics as follows: Intelligence tests an agent's capacity to accomplish objectives in a wide variety of environments. The challenge of knowing what intelligence is leads directly to attempts to emulate it in computers. For several decades, the term Artificial Intelligence has been around and has held very different connotations, based on technical development at the respective time [1].

The fact that significant tech corporations' marketing departments have seized the word AI for their purposes has not helped enhance transparency. If we apply the term strictly, then today, there is no artificial intelligence. In various conditions comparable to the one in which humans work, no current computer program can achieve objectives [2]. On the other hand, algorithms can perform specific, well-defined cognitive tasks at the (super) human level, such as playing a computer game or recognizing a dog's face in a picture. The AI group, therefore, distinguishes between Narrow AI and Strong AI. Like the above concept, we may describe Narrow AI as an agent's ability to achieve objectives in a (very) restricted range of environments. However, the realization of Powerful AI will possibly result in the most drastic changes in human history in culture and economy.

Today, the exact essence of these changes cannot be anticipated, and some of the several different imaginable possibilities are outside the reach of this article. We will discuss common claims propose against creating Powerful AI to support this point of view to the ownership of complexity. "The human brain is so extremely complex that we will never be able to reproduce it or mimic it (or at least not soon)." The human brain is enigmatic and incredibly complicated because it is by far the most sophisticated. Most of the brain's complexity is attributable to being a biological device that needs to satisfy many computer-irrelevant requirements: it needs to be utterly self-assembled in each phenotype, it undoubtedly has much needless complexity for historical evolutionary purposes (Bear in mind that nature is a tinkerer, not an inventor!), it's got to take care of a complex "hardware blend". It also has to get by with what little resources a biological body can produce. It is limited to "biotechnology" variants (i.e., nerve cells) that evolution has arisen for less intelligent animals to develop different scenarios. These entire criteria package be dropped for our purposes of developing Powerful AI, on the other side. We don't need to emulate a complete human being-the main algorithms that have its intelligence are entirely sufficient to reproduce it. To make this point simpler, consider the following analogy: Before discovering aerody-namics, people researched and admired birds for their ability to fly. The anatomical analysis will show that a bird is an incredibly complicated "machine" composed of or even only its wing, bones, organs, arteries, feathers, etc., and that there is no chance of replicating the wing of a bird down to the most delicate details of each feather's struc-ture. It is not that hard to build an airplane once the fundamental concept is under-stood. It turns out that most bird-specific complexities (such as flapping wings or delicately structured feathers) can happily happen dropped. Then, compared to the complexity of the brain as such, there is a second reason to believe that intelligence arises in a relatively simple way: the neocortex is the brain region that performs cogni-tive tasks such as recognizing images, understanding spoken language, controlling body movement, or thinking in general [3].

1.2 Machine Learning: An AI Subset

1.2.1 Five Essential Subsets for Artificial Intelligence

Artificial intelligence refers to a software or computer device's ability to mimic intel-ligence of humans (cognitive process). Straightforward meaning is disrupted, safe from experience, adapted to the latest results, and works like humans' exercises. Artificial Intelligence performs functions that are smart enough to create enormous precision, versatility, and efficiency for the whole system. Tech Chiefs are searching for a confident attitude to incorporate artificial intelligence technology in their or-ganizations to unblock and provide principles. For example, AI is used in the bank-ing sector, tourism sector, predicting stock prices, etc. Linguistics, bias, robotics, design, perception, natural language processing, decision-making, and other artifi-cial intelligence methods are all well-organized [4].

1.2.1.1 Machine Learning

ML is perhaps the most critical subset of AI in today's median enterprise. As explained in the Executive instruction for real-world AI, our modern analysis paper, which is regu-lated by the Harvard Business Review Analytic Services, ML is a full-scale invention that has been around for a long time. ML is a part of AI that allows machines to learn from statistics independently and implement learning without human arbitration. The cir-cumstances in which the suspension is protected in a large data set are challenged; AI is a go-to. "ML exceeds assumption at processing that details, extracting patterns from it in a small quantity of the time a human would take and distributing in any case out of reach awareness" quoted by Ingo Mierswa, president and founder of the RapidMiner

Data Science Network. ML powers possibility analysis, fraud detection, security conduct in Economic services, GPS-based forecast in travel and targeted marketing drive, to list a few examples. Much work is being carried out to improve the quality of software development. The work which is presented in this article is validated using various test processes [5, 6] and test techniques [7, 8].

1.2.1.2 Neural Network

The artificial intelligence system consists of many important components and neural network is one of the vital components that integrate the nervous system's technologies, combining cognitive science with techniques for performing tasks. The neural network is designed to replicate the human mind, which consists to an infinite number of neurons that are coded into a system or a computer. Combined, neural networks and deep learning handle complex tasks quickly while automating a vast portion of these tasks. NLTK is the sacred library of goals that are used in NLP. Ace all the modules, and you are immediately going to be a competent text analyzer. Pandas, NumPy, text blob, Matplotlib, word cloud, and other Python libraries are examples.

1.2.1.3 Deep Learning

There is a helpful example in an explanatory article by AI software company Pathmind: Think of many Russian dolls settling inside each other. Big data is a subset of machine learning, and artificial intelligence is a subset of AI, a common term for any software application that does something smart. Deep Learning uses alleged neuronal pathways that, according to one clarification offered by deep AI, learn from the production of the marked information presented during learning and use this response type to recognize which elements of the information are required to produce the correct output. The human brain can begin to process new, innocuous sources of information and effectively return accurate results when a sufficient number of models have been developed. For Amazon and Netflix, deep learning drives product and entertainment reviews. It functions in the framework of Google's algorithms for many applications, which are based on voice-controlled and recognition of images. Machine learning is unmistakably suitable for overcharging preventive maintenance systems because of its capacity to dissolve a bunch of high knowledge [9]. Malware assault is an extensive area of cybersecurity threats that impact networks and damage confidential data. The value of identifying and preventing a malware attack in time is the need of the hour [10]. Deep learning is implemented in various fields like recommender systems, music generation [11], diagnosing acute diseases, weather forecasting, agriculture sector, political tweet analysis etc.

1.2.1.4 Robotics

There is a helpful example in an explanatory article by AI software company Path-mind: This has emerged as an incredibly steamy artificial intelligence area. For the most part, a fascinating field of groundbreaking practice centers on creating and building robotics. Automation is a scientific discipline strengthened with computer science and engineering, electrical engineering, mechanical engineering, and many others in science and technology. It determines how robots are designed, made, controlled, and used. It oversees computing programs, intelligent outcomes, and data change for their regulation. Robots are routinely introduced to guide activities that could be difficult for humans to execute reliably. Big logistics tasks included a production line for the production of vehicles for NASA to move multiple objects in a vacuum. Besides, artificial intelligence researchers build robots using machine learning to set social communication standards [12].

1.2.1.5 Computer Vision

By marking the things in your dwelling with the native tongue and interpreted words, have you taken a chance at learning a new language? Since you see the terms, again and again, it is a good vocab creator by all accounts. The case with machines fueled by quantum computing is the same. However, by marking or categorizing different items, they learn to manage the repercussions or decode them at a great deal faster rate than individuals (similar to the robots shown in science fiction movies). By extending them to arithmetic computations, the OpenCV tool empowers image processing. Recall the elective course called "Fluffy Logic" in engineering days? This technique is used in image processing, making it much easier for deep learning experts to fuzzify or myste-rious readings that cannot be positioned in Yes/No or accurate brittle category. The OpenTLA is used to monitor video, wherein the moving object(s) is detected using a live feed from the camera.

1.3 Industry 4.0

1.3.1 Introduction

Industry 4.0 is the latest revolution in the industrial sector aimed at improving the 21st-century efficiency but then again also the versatility, compliance and durability of industrial structures. It helps the versatility, suppleness, and durability of indus-trial systems to capture real-time information from industrial organizations. The inno-vation in the Internet of Things (IoT) technologies empowers the gathering of factual data from industrial networks. Due to the progress of the Internet of Things (IoT) technologies, the online processing of captured statistics allows structures to be analyzed. Therefore, observing data gathered electronically helps many manufacturing challenges be handled quickly, such as system breakdowns or slow-downs, quality crises, flow disturbances etc. The main elements of Industry 4.0 are as shown in Fig. 1. Industries have to cope with many real-time manufacturing problems such as breakdown or slow-down equipment, efficiency crisis, disruptions of flows etc. Previous studies focused mostly on scheduling and restructuring schemes in conventional industrial systems to improve the framework of traditional industrial methods. Earlier projects focused on both planning and rescheduling systems to increase the efficiency of the system. Nevertheless, few works tackled system disturbance monitoring due to the system output's absence of actual data. However, owing to the deficiency of real-time details about the device operating, limited works tackled structure interruption monitoring. Besides, the remote and continuous control facilities have not been set up yet.

Also, the remote and continuous control facilities have not been set up yet. But many device disturbance control mechanisms in the Industry 4.0 scheme are available. These instruments hub on resource localization-related device disturbance, a disorganization platform in the Industry 4.0 system. These tools focus on device disturbance due to the localization of services or where a resource is uncomfortable. Therefore, a machine learning model is employed to create a resource localization forecast model by considering resource localization's actual scheduling tasks. Thus, globalization of services as natural resources when understanding the real scheduling functions in resource localization. Therefore, as original resource localization can be found through the IoT network from the economic structure, the tool allows service disturbance to be detected, risk, translation can be obtained through the IoT network from the industrial economy, and the tool enables usage disorder, risk, to be caught in real-time by correlating expected localization to the real one. Moreover, when compared characteristic localization to the actual one, the new tool is implemented in real-time Fog computing architecture. Besides, this new tool is executed in a Fog computing platform evolving as a cloud computing extension to provide acceptable latency for local processing support.

The experiments that are developing as an application of cloud computing supplies adequate bandwidth for shared storage support. In contrast to other computer outcomes, the empirical findings indicate our instrument's efficacy in classification accuracy and time complexity to more thoroughly monitor and track device interruption in real-time.

The fourth industrial revolution, generally called Technology 4.0, is the origin of emerging innovations in production. The industry 4.0 is focused on the comprehensive introduction of the Cyber-Physical System (CPS) and targets, through the processing and review of real-time data, to increase the competitiveness of 21st-century industrial technology dramatically. The innovation in the This transformation is primarily driven

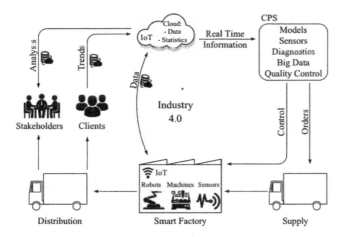

Fig. 1: Main elements of Industry 4.0.

by Internet of Things (IoT) technology, in which CPS components, devices, and objects are rendered "digital" by connecting them to the internet through pervasive sensors. Therefore, in many industrial areas, such as production, distribution, wholesaling, and pharmacy, IoT has been significant. Although related words such as Smart Production, Industrial IoT produced in China, are still used in Germany.

The core premise of Industry 4.0 is to bind the IoT and Services (IoT'S) system to the CPS, which interconnects the traditional and cyber realms. Thus, the gathering of real-time statistics analysis on CPS modules causes many industrial device problems to be solved. For example, the study of captured real-time information helps the African gold mine recognize a significant issue related to controlling one of its process phases. Fixing this dilemma has made it possible to increase the process's output by 3.7%, which translates to $20 million saved year after year.

In particular, prior work aims at providing either a robust type of analysis that can be the consequence of minor device failures and thereby stop rescheduling operations. Successful rescheduling schemes may lead to unanticipated circumstances such as computer breakdowns and cost-effectively review programs. With the advent of Industry 4.0, the monitoring of process interruption is becoming an essential research issue to enhance network performance. Thus, real-time monitoring of conditions facilitates a prompt response and aims to minimize, prevent, or prevent their effects. Real-time interruption control also makes it much more comfortable, if paired with an integrated and Just in Time (JIT) decision-making mechanism, to boost the network's quality, stability, and flexibility. Fog computing is also an evolving cloud computing extension that allows data near its source to be stored and is liable for business intelligence, device latency, etc. In the framework of Industry 4.0, it may also be very beneficial to implement Fog computing as a cloud infrastructure among

Industry 4.0 and Cloud settings to enhance the manufacturing system's capacity to avoid and respond to disturbances.

In this work, we discuss computer disruption control in Industry 4.0 built on the Fog computing architecture. We concentrate on the system disturbance associated to the globalization of assets in the production plant when there are several types. To do so, we create a resource localization projection model that considers the given document at any time during the entire setting up procedure in terms of resource localization. Therefore, by comparing modified system resources localization knowledge (human, instruments, etc.) to the expected one, we will forecast the probability of device disruption in real-time.

1.3.2 Related Work

This segment suggests a device disruption tool based on the Fog computing platform in Industry 4.0. We split the relevant workings into 2 fundamental classes: assignments that provide Industry 4.0 and processing domains with Fog computing implementation and works with device disturbance in the industrial sense.

1.3.2.1 Fog Computing used in Industry 4.0

Numerous efforts have recently signified to implement Fog computing as web services backing between Industry 4.0 and Cloud settings to execute the complex task quickly. How could Fog be used in Industry 4.0 as computational resources support to arrange for delay, security, and operational large data mining in real-time and in various industrial situations? Along with other industrial usage examples, they discuss the fundamental problems that the Fog will take advantage of, including mining, shipping, waste treatment industry, food industry, agriculture, etc. To include an intelligent computer infrastructure that involves cloud centers, gateways, Fog devices, edge devices, and sensing devices, implementing a Fog computing system in a logistics centre was investigated. Until solving it using genetic algorithms, the authors studied the cost-efficient application of Fog computation and proposed an integer programming model. To place a buyer node at the Fog layer in charge of forecasting future statistics of data such as energy consumption for IoT devices. They then researched Fog computing's role in improving the challenges of time and analysis. In practise, fog computing in the context of Industry 4.0 is still in its early stages of implementation. The research above looks at how to use fog computing to deal with the complexities of Industry 4.0. To the best of our knowledge, little research has been done to use fog computing to remotely and in real-time predict device disruption in Industry 4.0.

1.3.2.2 System Disruption in Industrial Systems

By considering such forms of disturbances, a layered disruption paradigm is pro posed to research and understand the impact of disruption in an entire industrial environment. Several production cases are then analyzed using both primary and secondary data through the proposed structure. The suggested framework seeks to have a stable initial timetable that can absorb the consequences of minor delays and thereby discourage activities from being rescheduled. Thus, multiple interruption patterns, e.g., system breakdowns, were considered in the initial scheduling. Correspondingly, several constructive disruption models have been developed by providing initial robust scheduling based on statistical patterns, deterministic expressions, Markov decision mechanisms, and so on. They may not, however, recommend tracking machine disruption in real-time. When determining the initial schedule, others only consider device disruption patterns and thereby prevent expensive rescheduling operations. Moreover, the majority of them concentrate on machine failures as a source of disruption, and no study has found a correlation between system disruption and unintended capital localization.

1.3.3 Tool to Predict the Disruption of System

This work presents an app that was introduced for structure disruption prediction in the Fog-computing environment. This approach is focused on an estimation method for resource utilization based on real task planning that we have produced at any time and across the entire planning process, assuming that we obtain accurate resource localization from the development execution system. As a result, by comparing actual localization to those expected by the model based on current tasks, the probability of device disruption can be assessed. We note that in that system, a resource relates to any entity that is needed to perform a given task and during the execution process, such as workers, machines, robots, etc. We first provide the Fog-based architectural design that we considered in our system before we continue.

1.3.3.1 Fog-based Architecture for Industry 4.0

This section suggests about Fog-enabled Industry 4.0 architecture with three primary levels to increase device efficiency and latency:
- **Industry 4.0:** It emphases on the production system, which includes sensors, actuators, devices, machinery, staff, and other "material." These things are self-contained, intelligent, and interconnected in order to complete tasks. Additionally, collected data is sent either to Fog computing for real-time data processing

and task management, or to the Internet Cloud for a more robust industrial data analytics analysis.

- **Fog computing:** It becomes a provision for web services that perform local real-time jobs such as rapid response delivery and offloading computing. In our case, based on predicted resource localization, the Fog is liable for threat prediction in the Industry 4.0 system. Therefore, it will support system superintendents to make the most appropriate choices such as rearranging the tasks, substituting the disastrous resource, initial periodic repairs etc.
- **Internet Cloud:** The Internet Cloud is mostly used to accomplish intricate jobs requiring enormous computing and storage capacity. In our architectural design, the Internet Cloud environment is answerable for creating our estimation model, which can then be used to predict real-time detection of resources by Fog computing.

1.4 Applications of AI and ML in Industry 4.0

1.4.1 Smart Manufacturing

Industry 4.0 (I4.0) includes a surplus of emerging innovations that impact manufacturing firms. Most research on this subject discusses the shrewd factory domain's impact, concentrating on the production schedule. However, comprehensive research on Industry 4.0 applications that enable technologies to generate life cycle processes is still lacking. The implementation of Industry 4.0 technologies may have a potential effect on the manufacturing companies' various strategies. Business processes may also be assisted by technology based on multiple applications, impacting both the customer and the entire value chain. The flow list includes the entire manufacturing process, from product design to assembly, operation, and, finally, recovery or disposal. As a result, all of the proven processes have an impact on the product during its lifecycle. Supply Chain Management (SCM) applies more accurately to part of these processes, i.e., supplier and customer management tasks are planning and handling material flows [13]. Smart manufacturing is an intelligent manufacturing process consisting of a physical image linked via a digital thread to its digital twin. The digital thread allows the physical world to interact and exchange data with its virtual reflection and digital twin. The digital twin will forecast, view, and assist the creation process with expert expertise. Intelligent (smart) growth provides excellent opportunities for all new technologies, such as cloud computing, the Internet of Things, big data analysis, and artificial intelligence. These methods were used to collect information in real-time, extract knowledge, and generate well-informed details [14].

1.4.2 Autonomous Vehicles and Machines

Without human intercession, an autonomous vehicle can control itself and carry out significant conditions through the ability to sense its surroundings. Autonomous vehicles are the way of the future, where humans no longer drive and are instead powered by computers. Making left-hand turns (right-hand turns in India) is dangerous for drivers because they must pass through oncoming traffic [15]. In the transport field, the advancement of artificial intelligence (AI) has stimulated the production and deployment of autonomous vehicles (AVs). It is essential to know how AI functions in AV systems to achieve complete automation. Existing researchers have produced great efforts to explore different aspects of the application of AI to AV production. The increased production of autonomous vehicles (AVs) has piqued global interest in recent years. AVs for transformative modern transportation systems are expected to address a number of long-standing transportation issues such as congestion, protection, parking, and energy conservation, among others. Many AV advances, it could be argued, have made considerable progress in bringing AVs into real-world applications in lab tests, closed-track tests, and public highway tests. Many shareholders, such as transportation authorities and IT behemoths, have reaped significant benefits from significant investments and promotions [16].

To solve difficult challenges in vision and motion planning, autonomous vehicles depend on machine learning. However, automotive software safety standards have not yet been completely established to address machine learning safety concerns such as interpretability, verification, and performance limitations [17]. The future of transportation is autonomous vehicles, and they are also projected to become a complete reality inside a span of next 10 years. The big automotive giants, such as Mercedes with Bosch, Toyota with Microsoft and Amazon, Audi with Huawei, etc., are struggling to changeover from traditional vehicles to autonomous vehicles. Consequently, this poses many autonomous cars opportunities as they can intensely shrink the accident frequency and increase effectiveness and trouble-free parking, optimum running time and helps in saving fuel. When all the vehicles can be coordinated together via the cloud, it also decreases traffic congestion. Significant research has been undertaken to model the idea of self-driving vehicles using deep learning and Real-Time route recommendations [18].

1.4.2.1 Autonomous Vehicle and its Challenges

- **Innovative techniques and sensors used in AV:** Autonomous vehicles focus on cameras on both sides. Blind-spot detection cameras with a super-wide focal point are available on some cars, giving the driver a broad view of what is behind the car. Even though they have accurate visuals, the camera lacks the ability to discern artefacts in low permeability conditions such as mist, rain, or evening time. Radar sensors can improve camera vision and self-driving vehicle position in this situation [19].

- **Autonomous vehicles in different systems, components and its challenges:** Current technology in autonomous vehicles is not well suited to be positioned in other climatic environments, such as a heavy downpour, snowfall, and so on. In the case of dips and continuing adjustments inroads, the system fails to consider this. A minor software failure can result in a disastrous situation for both the vehicle and the passenger [20].

- **Challenges for AV for Indian Scenario:** Major remote vehicle players do not see India as an upright marketplace for autonomous automobiles. Another obstacle to autonomous vehicles is the low road and transportation infrastructure in the country.

1.4.3 Quality Control

In reality, Industry 4.0 is one of the key and most discussed topics in academia and practice, providing enormous benefits for companies and new opportunities for a range of applications. In addition to this trend, to remain competitive and satisfy consumers' ever-increasing demands, today's manufacturing companies have to deliver products of the highest quality. Therefore, focusing on quality management is essential for any company and a key to sustainable economic growth. Via its values, Industry 4.0 provides promising prospects for quality management (Smart Plant, Cyber-Physical Infrastructure, Internet of Things and Services) [21].

The aim of quality mitigation strategies and procedures is to ensure that the quality specifications are met.

- **SPC:** The SPC (Statistical Process Control) is a basic quality assurance tool that uses sampling-based leverage to accurately track the quality performance of industrial processes and the proper timing of appropriate procedures.

- **Audit:** One of the audit's main goals is to ensure that the company's practices adhere to best practices as well as the guidelines established by the quality system. As a result, auditing is one of the most important methods for implementing quality control.

- **QTC:** The QTC (Total Quality Control) concept is unique in that it applies quality control to the entire enterprise, as well as the product's entire life cycle. This is achieved by including all branches of the company, thereby going beyond the traditional growth boundaries [22].

1.4.4 Predictive Maintenance

Historical data based on evidence, models, and domains makes up preventive modelling. To predict upcoming failures, it can use statistical or machine learning forecasting methods trends, activity patterns, and correlations to improve the maintenance operation's decision-making process, mainly avoiding downtime. The industrial revolution is something that Industry 4.0 is really involved in. Every day, both machines and managers must make decisions requiring a large amount of data and customization in the manufacturing process. One of the most difficult issues is predicting asset repair at a specific time in the future [23].

To survive in today's highly competitive market climate, modern companies need productivity and comfort in managing the entire Product Life Cycle (PLC). In smart manufacturing environments, IoT paves the way for creative predictive maintenance techniques, and these sophisticated technology produce large amounts of industrial data. To make the most of this big data, trend-oriented predictive maintenance activities are carried out given the recent state of the machinery to foil catastrophes. Advanced enabling technology and wireless technologies enable innovative ways to track machine standing and actions in microscopic detail. Even if predictive maintenance enables increased business benefits and smart machining services, huge information from multiple technologies must be managed effectively for compatibility through standards for data blending and transformation [24].

Classifying the approaches used for predictions

- **Physical model-based:** The most important aspect is mathematical modelling with reflexes in the state of a variable, which necessitates condition correctness and failure assessment, as well as statistical methods to limit these indices.
- **Knowledge-based:** Methods that reduce the complexity of a physical model, such as expert systems or fuzzy logic, are used as a hybrid approach for this purpose.
- **Data-driven:** The most commonly used models in the current evolution of PdM solutions are statistics-driven, pattern recognition or artificial intelligence (AI), and models based on machine learning algorithms.

1.4.5 Demand Prediction

The growing demand for data and the explosion in the number of sensing devices which, in terms of communication, battery and computing power, could be highly limited. Supply and demand are two basic principles for vendors and consumers. Predicting demand is crucial for companies to be able to make preparations for procuring and storing necessary products. Through the pressure of Industry 4.0, manufacturing companies are transforming themselves through the Internet of Things towards digitization and improving their supply chain activities (IoT).

Also, manufacturing industries transformed with AI and ML techniques by integrating greater visibility, versatility and operational skills into the supply chain phase. It requires rigorous demand forecasting and, by formal scenario analysis, enhanced decision-making. It also facilitates the process of inventory optimization by using methods of statistical modelling. Thus, it performs the amount of inventory stock and the study of missed sales scenarios [25].

1.4.6 Chat Bots

A Chat bot (or Chatterbox) is a software application that communicates with humans. It is a virtual assistant capable of answering several questions and providing accurate answers. Chat bots have become ever more popular in a variety of fields in recent years, including health care, marketing, education, support systems, cultural heritage, entertainment, and many others. Chat bots for industrial solutions and analysis have been developed by a number of major companies, including Apple Siri, Microsoft Cortana, Facebook M, and IBM Watson. These are just a few among the most widely used applications. One of Chat bot's applications is in the field of education.

1.5 New Challenges in Industry 4.0

AI is pushing the Revolution of Industry 4.0 through various sectors. Machine learning is behind this revolution as it forms the core for analyzing colossal data in real-time. The problems surrounding industry 4.0 need to concentrate on implementing emerging technology by improving mechanical and robotic processes and optimizing other areas like manufacturing, customer support, administration etc., through the use of research schemes and technological advancements.

1. **The hurdles faced by small and medium-sized businesses while adopting Industry 4.0:** Analysis of data in real-time for decision making: Gathering data through various means, storing the data and analyzing the data in real-time will help in the strategic decision-making process. Know your customer: Getting to know the customer's ever-changing needs is the key to attracting the customers. The industries should anticipate customers' needs through novel marketing policies and accordingly modify their approach. Adapting to new technology: Industries must embrace the latest technologies to benefit the company's growth in the long run. It is always difficult to adjust to new technology. Nevertheless, with the right human resources, it can be easily achieved.

2. **The Smart Sensors technology:** The operation of smart sensors focuses on providing computers with the ability to see, track and communicate intelligently. This feature is distinguished by monitoring manufacturing systems,

which allows for identifying faults or defects, making operations more productive and successful.

3. **The challenge of digitalization and automation:** Thanks to IoT, processes would be managed from any form of Smartphone device, resulting in a strategic edge vis-à-vis suppliers that use more traditional methods. The idea of versatility is essential to improving times, lowering costs and promoting customer-manufacturer contact at all times.

4. **Security:** Network interconnection is a key feature of commercial digitalization, but it also poses a security issue with respect to data privacy. These data must be shielded from overt, external hacking threats and against accidental data breaches, such as a consequence of a mistake or lack of proficiency by the member of staff.

5. **Testability:** Until deployment, each new system or system update must be evaluated in an industrial environment to check the security and consistent response in several circumstances. Monitoring was and still is a vital part of the transition process, even in the age of independent OT and IT systems. When an enterprise is completely developed and digitized, analysis becomes more complex now than before.

6. **Implementing artificial intelligence:** To correctly implement and conduct artificial intelligence applications, most algorithms require a training process using a standard data set. However, collecting such data is another significant task. Not only is it necessary to capture a reasonable amount of data specific to learning, but it must also be seen that this data preserves all critical system states enough so that the whole system can be adequately monitored.

7. **Distributed, Disconnected and Legacy Systems:** The market for manufacturing equipment is fragmented and is comprised of many national and international players. In the manufacturing industry, a common problem is the lack of system interoperability. Suppliers have a variety of technologies and components that are unfamiliar with other networks. Standards and frameworks for industrial IoT are insufficient. The correct approach to this problem is a network of players that can recommend suitable apparatus that use classic set of laws and frameworks to interact with ERP, PLC/SCADA, MES and systems.

8. **Demand for Real-Time Response:** In Industry 4.0, condition surveillance is a regular use case where the asset's real-time data is analyzed using streaming monitoring to ensure maximum performance. Several latency-sensitive applications in production, such as predictive protection or analytical quality, call for an ultra-fast reaction. The framework cannot linger for the round trip to the cloud for these applications to carry out data analysis and get actionable insights. The decision needs to be made in real time, with strong action in a few milliseconds. Edge tech is going to make a difference here. For faster response, processing data locally near the data source becomes more effective.

1.6 New Opportunities in Industry 4.0

Digitalization of manufacturing and computerization of manufacturing will signal the transformation. Significant components that forever change the industry are described below. Smart and automated systems aided by big data and deep learning, for example, will make construction and shipping smoother and far more beneficial now than ever. For smart manufacturing, the main implementation of Industry 4.0 will be Industry 4.0; however, it may also lead to smart docks with autonomous cranes effective for loading freight into ships without manual assistances. If vehicles, trains and aircraft becoming more autonomous, reducing dependence on passengers, pilots and drivers, the transportation would also have a major effect. There could also be smart towns right around the corner.

- **Big Data and Analysis:** Big data analytics can analyze and gain valuable knowledge from large amounts of information. Big data analytics are continuously developing and soon will be used more intensely in Industry 4.0.
- **Internet of Things (IoT) platforms:** All devices that can gather data, send it over the internet, and connect with other devices are referred to by the Internet of Things. All examples of IoT devices include smart refrigerators, lamps, and toasters.
- **Blockchain Technology:** Blockchain is a technology that emerged from the popular digital currency called as Bit coin. This technology contains knowledge being stored in the form of "blocks" on universal public ledgers that are linked to each other on a "chain" and are checked by "miners."
- **3D technology:** 3D technology is an evolving industry that is getting adopted by businesses today, including 3D modelling, 3D printing, 3D rendering, 3D display, and so on, to enhance the shopping experience and to simplify the working process.
- **Smart factory:** Mighty, safe and cost-effective factories rely on sophisticated technologies such as robotics, big data analytics, cloud computing, robust cyber security and smart sensors.
- **Location detection technologies:** Such technologies identify your location and are often seen on mobile devices. With location tracking technology, you can share your locality with trusted individuals.
- **Customer Interaction and Profiling:** Based on such identifiers, this technology involves grouping clients into classes. Hobbies, age, place, interests, etc., may include such identifiers.
- **Advanced Human-Machine Interfaces:** Advanced human-computer interfaces are machine interfaces that endow with illustration information about the machine's functions to allow operators to understand how the device operates in real-time.
- **Waste Management System:** A lot of research has been done on waste management. Waste disposal refers to the elimination of waste through recycling

and ground filling. Deep learning and the Internet of Things (IoT) offer an efficient solution for classifying and tracking data in real-time [26]. Using these and other methods, we can separate waste and handle wet, dry, and e-waste virtually [27] [28].

1.7 Conclusion

Industry 4.0 relates to the implementation of innovation and the advancement of expertise, services and high-tech infrastructure to develop manufacturing plants. A variety of industrial technologies are being developed and implemented using AI and ML. Different categories of businesses and analysis outputs are expected to operate on Industry 4.0 systems, including the use and application of AI, ML, Big Data and the Internet of Things (IoT). There is also an immediate need to develop future-proof AI and ML software, services, architectures and proof-of-concept. In the Industry 4.0 ecosystem, information gathered by sensing devices involves machine learning and data analysis techniques. As a result, companies face the new possibilities and obstacles, a statistical analysis using computer software capable of identifying trends in the analyzed data from the same laws that can be used to formulate forecasts. Innovative intelligence and machine learning approaches can help the decision of architecture and offer multiple advantages such as faster decision-making, preservation of business information, saving man-hour, higher computational accuracy, and speed.

References

[1] Niewiadomski, P., Stachowiak, A., & Pawlak, N. (2019). Knowledge on IT Tools Based on AI Maturity–Industry 4.0 Perspective. Procedia Manufacturing, 39, 574–582.

[2] de Paula Ferreira, W., Armellini, F., & De Santa-Eulalia, L. A. (2020). Simulation in industry 4.0: A state-of-the-art review. Computers & Industrial Engineering, 106868.

[3] Bench-Capon, T. J., & Dunne, P. E. (2007). Argumentation in artificial intelligence. Artificial intelligence, 171(10–15), 619–641.

[4] Tantawi, K. H., Sokolov, A., & Tantawi, O. (2019, December). Advances in industrial robotics: From industry 3.0 automation to industry 4.0 collaboration. In 2019 4th Technology Innovation Management and Engineering Science International Conference (TIMES-iCON) (pp. 1–4). IEEE.

[5] Naresh E. and Vijaya Kumar B.P. 2018. Innovative Approaches in Pair Programming to Enhance the Quality of Software Development. Int. J. Inf. Comm. Technol. Hum. Dev. 10, 2 (April 2018), 42–53. DOI:https://doi.org/10.4018/IJICTHD.2018040104.

[6] E. Naresh, (2020). Design and Development of Novel Techniques for Cost-Effectiveness with Assured Quality in Software Development. Ph.D thesis, Jain University. http://hdl.handle.net/10603/288589.

[7]   Naresh, E., Kumar, B. P. Vijaya Kumar., Niranjanamurthy, M., & Nigam, B. (2019). Challenges and issues in test process management. Journal of Computational and Theoretical Nanoscience, 16(9),3744–3747.

[8]   Naresh, E., Vijaya Kumar, B. P., & Naik, M. D. (2019). Survey on test generation using machine learning technique. International Journal of Recent Technology and Engineering, 7(6),562–566.

[9]   Olson, R. S., Sipper, M., La Cava, W., Tartarone, S., Vitale, S., Fu, W., . . . & Moore, J. H. (2018). A system for accessible artificial intelligence. In Genetic programming theory and practice XV (pp. 121–134). Springer, Cham.

[10]   Devika A, Aravind P A, Aatish K, Abishek P & Rajarajeswari S. (2019). Windows Based Malware Prediction System using Deep Learning Techniques. Journal of Computational Information Systems (JOF-CIS) Pg No: 339–354 Volume-15 Issue-3, 2019 ISSN: 1553-9105

[11]   Prashant Krishnan, V., Rajarajeswari, S., Krishnamohan, V., Sheel, V. C., & Deepak, R. (2020). Music Generation Using Deep Learning Techniques. Journal of Computational and Theoretical Nanoscience, 17(9–10), 3983–3987.

[12]   Linde, H., KGaA, M., & Schweizer, I. (2019). A White Paper on the Future of Artificial Intelligence.

[13]   Zheng, T., Ardolino, M., Bacchetti, A., & Perona, M. (2021). The applications of Industry 4.0 technologies in manufacturing context: a systematic literature review. International Journal of Production Research, 59(6),1922–1954.

[14]   Evjemo, L. D., Gjerstad, T., Grøtli, E. I., & Sziebig, G. (2020). Trends in Smart Manufacturing: Role of Humans and Industrial Robots in Smart Factories. Current Robotics Reports, 1(2), 35–41.

[15]   Prasath, G. S., Poopathi, M. R., Sarvesh, P., & Samuel, A. (2020, August). Application of Machine learning Algorithms in Autonomous Vehicles Navigation System. In IOP Conference Series: Materials Science and Engineering (Vol. 912, No. 6, p. 062028). IOP Publishing.

[16]   Ma, Y., Wang, Z., Yang, H., & Yang, L. (2020). Artificial intelligence applications in the development of autonomous vehicles: a survey. IEEE/CAA Journal of Automatica Sinica, 7(2), 315–329.

[17]   Mohseni, S., Pitale, M., Singh, V., & Wang, Z. (2019). Practical solutions for machine learning safety in autonomous vehicles. arXiv preprint arXiv:1912.09630.

[18]   Rajarajeswari, S., Reddy, S. R. P., & Venkatesh, A. (2013). Improving Real-Time Route Suggestions in Automotive Navigation Systems using Vehicle Cluster Behavior. International Journal of Computer Applications, 76(5).

[19]   Sales, D., Correa, D., Osório, F. S., & Wolf, D. F. (2012, September). 3d vision-based autonomous navigation system using ann and kinect sensor. In International Conference on Engineering Applications of Neural Networks (pp. 305–314). Springer, Berlin, Heidelberg.

[20]   Chen, L., Li, Q., Li, M., Zhang, L., & Mao, Q. (2012). Design of a multi-sensor cooperation travel environment perception system for autonomous vehicle. Sensors, 12(9),12386–12404.

[21]   Foidl, H., & Felderer, M. (2015, November). Research challenges of industry 4.0 for quality management. In International conference on enterprise resource planning systems (pp. 121–137). Springer, Cham.

[22]   Illés, B., Tamás, P., Dobos, P., & Skapinyecz, R. (2017). New challenges for quality assurance of manufacturing processes in industry 4.0. In Solid State Phenomena (Vol. 261, pp. 481–486). Trans Tech Publications Ltd.

[23]   Zonta, T., da Costa, C. A., da Rosa Righi, R., de Lima, M. J., da Trindade, E. S., & Li, G. P. (2020). Predictive maintenance in the Industry 4.0: A systematic literature review. Computers & Industrial Engineering, 106889.

[24] Paolanti, M., Romeo, L., Felicetti, A., Mancini, A., Frontoni, E., & Loncarski, J. (2018, July). Machine learning approach for predictive maintenance in industry 4.0. In 2018 14th IEEE/ASME International Conference on Mechatronic and Embedded Systems and Applications (MESA) (pp. 1–6). IEEE.

[25] Dhanabalan, T., & Sathish, A. (2018). Transforming Indian industries through artificial intelligence and robotics in industry 4.0. International Journal of Mechanical Engineering and Technology, 9(10),835–845.

[26] Gayatri M, S. Rajarajeswari. (2020). Survey Paper on Segregation of Waste Management System. Gedrag & Organisatie Review, Pg No: 249–252 Volume-33 Issue-03, 2020 ISSN:0921-5077

[27] Rahman, M. W., Islam, R., Hasan, A., Bithi, N. I., Hasan, M. M., & Rahman, M. M. (2020). Intelligent waste management system using deep learning with IoT. Journal of King Saud University-Computer and Information Sciences.

[28] Hiremath, T., & Rajarajeswari, S. (2019). A Survey on Existing Convolutional Neural Networks and Waste Management Techniques and an Approach to Solve Waste Classification Problem Using Neural Networks. In Emerging Research in Computing, Information, Communication and Applications (pp. 45–64). Springer, Singapore.

Siba Kumar Panda, Konasagar Achyut, Swati K. Kulkarni

Industry 4.0: A Road Map to Custom Register Transfer Level Design

Abstract: Time to market with an agile strategic approach for the development and production of products has led mankind to deliberately think or work in more anticipated way. One of the most recurring ways of production in today's industry is through automating almost whole development procedures. If we consider the expansion in technology from monolithic age till yesteryears, farming methods to usage of steam engines and from usage of electrical lines mechanizing almost every corner of the world to the introduction of computers has made a great effect on implementing man power on-field. Industry 4.0 is the name derived considering all the system-based mechanizations of work in the production field by usage of automation methodologies. When we consider automation, the major work here done is through the processor of any host controlling system. Processor is any such integrated circuit with billions of transistors in one single semiconductor fabric chip doing almost each and every complex work which a man can barely do. These complex works can be described by human through programming the processor chips with standardized languages which a chip will be able to understand before implementing over a fabric. So, now the time is of growing semiconductor industry. Technology is just being developed exponentially from past years just because of the capability of the processor chips being developed in the industry. This chapter focus on fore as well as background of the involvement of the semiconductor in industry 4.0 revolution as the digitization of non-human objects in the current generation days is at pinnacle. VLSI technology is the core of any IC development hub, thus the strategies that can be followed from base of VLSI designing, verifying IPs and finally taking forward the chip designing notion towards the fabrication industries are well described in the following sections. From the RTL designing of micro level architectures to the implementing of netlist over the fabric with challenging considerations such as power, static timing analysis, temperatures etc. are also equally important. This can be well understood under this chapter with step-by-step approach from RTL design to fabs with wide simulation practices.

Keywords: register transfer level (RTL), VLSI, IP, netlist, synthesis, static timing analysis, IIOT, AI, HDL

Siba Kumar Panda, Mobiveil Technologies India Pvt Ltd, Chennai, India,
e-mail: panda.sibakumar08vssut@gmail.com
Konasagar Achyut, J.B Institute of Engineering and Technology, Hyderabad, India
Swati K. Kulkarni, Dept. of Applied Electronics, Gulburga University, Karnataka, India

https://doi.org/10.1515/9783110725490-002

2.1 Introduction

In the recent studies, A. Ghaffari et al.[1] describes the impact of Convolutional Neural Networks (CNNs) on our society. This work describes the design and implementation of a general framework for developing a convolutional neural network on FPGAs. The authors described the CNN2Gate machine learning algorithm. This algorithm helps to extract the computational flow of layers, weight, and biases for fixed-point quantization. The author uses the popular Python Library. They also verified the functionality of the CNN2Gate machine learning algorithm on AlexNet and Arria 10 FPGA. Alasdair Gilchrist et al.[2] explained about new and innovative business models that arise from the IIoT as these are hugely attractive to business executives. He also explained key opportunities, benefits, the advantages and features of IIOT.A. Mahapatra et al. [3] proposes a method to optimize the RTL code. In this paper, author introduces a new RTL to C compiler called RTLcompiler2C.This complier will help to translate RTL IP. Author uses multi-dimensional array concept which helps expanding the search space as modern HLS tools allow to fuse them into single array. K. Nosalska et al. propose a conceptual framework [4] for Marketing in Industry 4.0. They derived the guidelines for designing strategies to implement Industry 4.0.Thiswork explores a better understanding of the Industry 4.0 singularity in terms of changes in the area of marketing in general and in industrial markets in particular. Here author explains 4 pillars of Industry 4.0 as Cooperation, Conversation Co-creation, Cognitivity and Connectivity. G. Karacay et al. explained how Industry 4.0 is growing, what are the factors associated with it [5]. Industry 4.0 era necessitates all employees, even workers with low-skilled jobs, to have a collection of ICT skills. However, Industry 4.0 requires essential employee skill sets to entail more than core skills; indeed, for successful execution of hard-skills, employees should have soft-skills as collaboration, communication and autonomy for being able to execute their jobs in hybrid operating systems. Y. Lu et al. [6] conducts a comprehensive review on Industry 4.0. The author has examined the existing literature in all of the databases within the Web of Science. The authors reviewed the content, scope, and findings of Industry 4.0. They explained the history of Industry 4.0 and the framework of interoperability. Y. Lu et al. [7] explained the exact mapping of industry 4.0 and categorized the papers as per the top-ten and other most frequently-cited industrial integration articles, Findings, publications in the research category of industrial automation and process. The major limitation of this study is that only SCI/SSCI database is adopted.B.YU et al. [8] discussed some key process technology and VLSI design co-optimization issues in nanometer technology. Authors also described the key challenge comes from lithography limits and manufacturability. Another key nanometer IC challenge comes from reliability, which usually refers to how robust a chip is after manufacturing. To overcome these issues, authors suggest full-chip modelling and CAD tools. The authors did survey on recent developments in design for manufacturability and reliability in extreme-scaling VLSI, including challenges, solutions, results and future research directions. Y. Ma et al. presented a scalable solution that integrates the flexibility of high-level synthesis and the optimization of an RTL implementation [9]. Here the authors described a brief overview of operations in a typical CNN model and the practical challenges in the FPGA implementation. The authors have implemented and verified the proposed CNN algorithms on different FPGA boards like Altera Stratix-V GXA7 FPGAs. They found the optimized and improved performance of the CNN module on FPGA. According to C. Zhang et al. [10], various researchers have already proposed a Convolutional neural network (CNN) and its implementation on an FPGA platform. This paper explains various optimization techniques. The authors achieved optimize CNN's computation and memory access. The suggested method is well suited for each layer. S. L. Chu et al. [11] proposed a novel design methodology, called data-oriented methodology using bluespec System Verilog and the corresponding tools. The proposed methodology simplifies the work by dividing the complex data paths into different individual blocks. Giovanni De Micheli [12] suggested various manufacturing technologies by surveying the design requirements and solutions for diverse systems and addresses design technologies. This work mainly focused on the semiconductor industry and post-silicon technology. L Beningnad H Foster et al. [13] describes the basics of simple RTL methodologies. EDA tools are capable to handle the challenges of the RTL. Still, verification variables and synthesizable variables that have emerged may cause the problem for semantic mapping. The authors suggest the subset of Verilog. This proposed subset of Verilog helps the HDL designer in terms of Tool Development cost and project schedule. K. R. G. Da Silva et al. [14] proposes a verification methodology (VeriSC2). Here authors do not consider test benches instead they proposed a new methodology called VeriSC2which explains the implementation of System-C Library, with tool support. A case study from a MPEG-4 decoder design is used to show the effectiveness of this approach. Case studies show a significant increase in design productivity with the used methodology. V. P. Shmerko et al. [15] describes the Malyugin's theorems which is a new approach in Logical Control, VLSI Design and data structures for new technologies. Author explained that, this theorem is effective for linearizing arithmetical forms of representation of not only Boolean functions but also multivalued logical functions in arithmetical logic and modern CAD IC tools. E. Takeda et al. [16] explained the new trend in VLSI. At the beginning of the 21st century, giga-scale memories and sophisticated processors with device dimensions as small as 0.1 pm or less, could appear. Revolution in the VLSI industry brought the paradigm shift. In this paper, the authors had reviewed and discussed the VLSI reliability issues concerning philosophical and practical differences between Japanese and American IC industries. Hirose et al. [17] explained the future of the VLSI domain of submicron technology. Here the author had discussed the VLSI process, future trends of VLSI device architecture, advanced chemical vapor deposition technology and advanced etching. The author focused on future trends in the silicon industry in terms of lithography and cryogenic process. K. Goser et al. [18] describes VLSI technology for Artificial Neural Network. According to the author's earlier implementation of the neural network, researchers were dependent only on computer simulation or dedicated computers. Now they can implement neural networks on microprocessors and microelectronic components because of the advancements in ICs technology. G. Rabbat et al. [19] explained the three aspects of the VLSI domain. In this paper, the author mentioned the impact of AI and expert systems on VLSI. The author explained the importance of AI in the VLSI industry. AI will be applicable at every stage of the semiconductor process interconnect technology, packaging, design techniques, and system architecture. Don Mills and Clifford E. Cummings [20] explained Verilog coding simulation and synthesis mismatch with suitable examples. The author suggests the guidelines to overcome simulation-synthesis mismatch. It is very difficult to find out the mismatch for complex design. To detect the mismatch in the code, the designer has to pass all possible combinations for testing and it is a tedious process. The author suggests a solution that avoids this problem. This fundamental concept shows a roadmap to the RTL design engineers for efficient RTL design in semiconductor industry.

The above discussion motivates us to understand the key aspects of Industry 4.0 precisely semiconductor industry. Considering the system specification to develop the efficient RTL code, then followed by simulation, synthesis, timing analysis, verification as well as fabrication of chip. Finally, it is believed that the presented chapter may establish a new route for imminent research.

The organization of this chapter is as follows: Section 1 will help in understanding the various state-of-the-art-techniques. Section 2 explains in brief about Fourth industrial revolution in Semiconductor engineering. Section 3 describes about Market trends and Challenges in Industry 4.0. Section 4 explains about the Step-by-Step approach from RTL design to Fabs. The Synthesized Modelling approach for Custom Chip design is described in Section 5.This will help in understanding the paradigm in order to get a synthesizable module. Section 6 focuses on the Challenges in Industry 4.0. Section 7 explains about the various applications of Industry 4.0. Section 8 deals with various Simulation Practices in Industry 4.0.In Section 9, conclusions are furnished.

2.2 Fourth Industrial Revolution in Semiconductor Engineering

Progressing through the go to market with respect to the updated technology and industrial revolution, we need to adopt from time to time and enhance our skills too. We being at the center of fourth industrial Insurgency that is Industry 4.0 [21–22],there is a lot of demand for exploration as well as innovations. The technological advancements involving VLSI Computing [23], quantum computing, artificial intelligence, Machine Learning (ML), the Internet of Things (IoT), the Industrial Internet of things (IIoT) [2] and Chip designs (RTL design) in the semiconductor industry become very popular and plays an important role for variety of applications. This new epoch always motivates researchers and engineers to think, predict and judge in an innovative way. A wide application of Industry 4.0 in semiconductor engineering is presented in Fig. 1. It focuses on the product development in terms of Integrated Circuits. This design can be done with the help of pure RTL code in digital platform. This is described through any of the Hardware description language either in Verilog or VHDL. Writing the RTL with respect to the specifications by looking into functionality is very important. Following through the simulation, synthesis as well as timing analysis, the development of chip will undergo the process of verification approach.

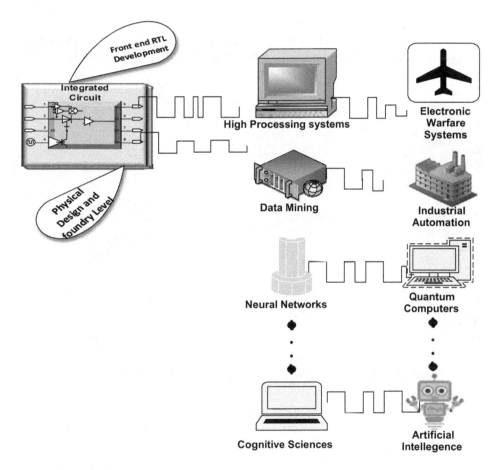

Fig. 1: Usage of IC in various fields for Industry 4.0.

This will mainly focus on the different techniques involved for custom RTL design with respect to semiconductor industry. In other hand of semiconductor and cutting-edge materials science industries play a critical role in the realization of Industry 4.0 due to manufacture of various computing devices using updated technologies. In addition, by espousing Industry 4.0 manufacturing strategies, the semiconductor industry is set to accelerate the pace of change by enabling greater development of the technologies such as micro-electromechanical systems and sensors, light emitting diode (LED) etc. The high-performance ICs Fig. 2, are the results of the advanced semiconductor materials where we can see the many fruitful research works are also going on. Chip industries [12] are digitizing their manufacturing environments in order to retain competitiveness and efficiency. The next generation technologies are pretty enabled with the skilled, vertical integration systems. The executions and machineries significantly diminish cycle times that also improves the throughput. Despite the advantages of digitization, there are many elements to take into consideration. By explaining the chip concept in terms of a smartphone, a defect in chip if found at the end step, it's a great loss to the manufacturer so, it should be found at early stage.

Fig. 2: Integrated Circuit pictorial representation.

2.3 Market Trends and Challenges in Industry 4.0

This section focuses on the history and market trends in Industry 4.0, specifically in semiconductor industry. Furthermore, this section describes the progression of Industry 4.0 and its future aspect too. In international usage, the Industrial Internet of Things, smart factory, and intelligent systems are used for the development of high-end systems. Basically, Industry 4.0 [4–7] states the global trend towards the digitization and networking of the industrial products. It is very important for manufacturers to use the newest technologies for their own production process and also on the other to put these technologies and products on the market. As supplier's viewpoint, the association among the information or communication knowledge and standard high-tech approaches mostly helps the growth of production process. As far as market dynamics and complexity are concerned, the expansion of industry has a direct interconnection.

To encounter the requirements of suppliers and manufacturers, company needs strategies that turn on both standpoints. The semiconductor industry like AMD, Analog devices follows the dual strategy by which they transform their own products in to full automations. The field of Science and engineering has evolved very fast and the developments are as per the market needs and demands. As progressing with the transformations, we are seeing some legal challenges highlighted as below.

– Deficiency of knowledge in the technology
– Uncertainty about the profits of technology investments for specific product
– Knowledge gap in customer demand about new products &business models as far as industry 4.0 dream is concerned
– Shortage of resources
– Requirement of efficient technology investments
– Requirements for proper scheduling of a new product launch
– Collaboration with specific partners
– Communication gap in the benefits of the Industry 4.0 revolution

Seeing the synchronization between design principles and supportive technologies, integrated business methods can be evaluated as the most important design principles. The effective implementation of Industry 4.0, emphasis on the involvement and refinement of the smart products as well as processes. The core technologies like VLSI, especially front-end designs are key to the semiconductor industries. The front-end design involves the step from custom RTL design to GDS-II are one part of chip designs. Then the design is subjected to verification process by injecting various test cases. The verification process is done by verification engineers, whereas the design is done by RTL design engineers in industry. Due to the enhancements in technologies, new business models, far-off services and continuous production operations are stimulated. Industries mainly focus on the perfect systems, not the single components of the systems separately. The mounting IoT, IIOT, Artificial intelligence, Machine learning, deep learning and the arrival of Industry 4.0 will create a world which is allied and united at each level. Currently companies are capitalizing huge amounts in optimizing different types of technological systems. Actually, this produces enormous need for innovations and explorations in the basic components in various types of industry like semiconductor industry. Hence, we can predict a high growth market for these components in future.

2.4 Step by Step Approach from RTL Design to Fabs- A Quick Guide

The current industry's thematic description is all the radix of digitization of every possible steps of the factory procedures in order to automate and replacing more and more human interacting with machines upon with machine interacting with machines. This in return has a great impact on time to market strategic approach of any company's revenue profit making practice. When we come to term digitization, especially when industry 4.0 has been standardized, has led the semiconductor industry lashing with many challenges towards them. Internet of things (IoT), robotics, deep learning, data mining, neural networks etc. have a common implication among all that is, the data has to be processed on a real time application. Which in turn can also be expressed as corner of all engineering is interconnected to the core of any processing system, nothing but an integrated chip (IC). Those processing systems are relevant to any of one such industry 4.0 application or any industrial revolution superseding. Supposedly, any company taken up a project to produce signed off chip having an application of data mining has its subsequent teams responsible for generating designs which tends to carry clusters of data understandable by a processor chip. During the development stage of in front end VLSI [8–12], design trade-offs have to be followed as given by foundries, this makes sure that the required chip production is not out bounding any constrained metric. Automatic assertion generation is one among those strategies to build a data mining-based RTL design. Among these development stages RTL designing is one such prominent period for any chip development to be signed-off by any manufacturer. For this, well understandable in addition to smarter approach is required to get through RTL designing stage because level of optimization in every succeeding stage is considered to be as high yielding output vessel for an end market, Fig. 3. The following section allows us to understand how any custom RTL designing can be followed with a modest change in standard market-based RTL design approach, also helping us to understand what really cooks at the back-end/physical design stage. Here we present you the steps gradually taking you from the coming up with ideas to the putting those ideas in a HDL, to the generation of final file from the code-based design to the implementation of code-based files over the physical fabrics present in the real dimensional world. We can observe that the cluster of wide range of applications based on different background of subjects from industry 4.0 revolution, all depending on the working of the main brain of their computing systems that is, nothing but an integrated circuit chip which is embedded with all the required programming instructions and also for the real time-based sensors conjoined with the processor chip/integrated chip. For this development, the first and the foremost step during chip designing is about idealizing about the required logic to perform such operation. These operations are well related to industry 4.0, because digitizing is a metaphor for this revolution. Thus, digitizing means more and more dependency over a processor chip. The following sub sections basically allows you to know exactly how a chip is designed in the VLSI/Semiconductor industry, front-end as well as in the back-end countenance.

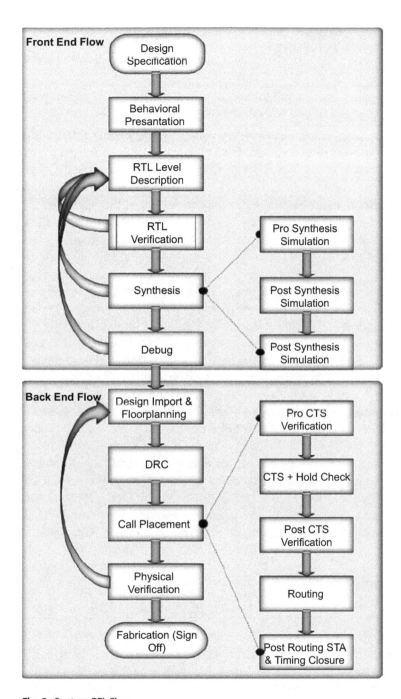

Fig. 3: Custom RTL Flow.

2.4.1 Design Specifications and Identification of Required Architecture

The basic opinion about the development or any changes in the functionality has to be specified among all the teams involved in the development of the same. The type of chip is dependent on the required application in the final. Basically, there are 3 kinds of development level chips, Field Programmable Gate Array [10], Application Specific Integrated Circuit (ASIC) and System on Chip (SoC). These are used based on the requirement of operations. When RTL or HDL designs are still in the development stage then FPGA boards or development kits are used so that the design can be altered or modified as required. Before reaching the extent of this point, design specifications are well stated before moving to the next stage. Design specification is a common step for all types of chip development. Unless you start laying out the specifications of your desired design among your team you would be running more towards perplexed way of creating design logics. Specifications are the high prioritized requirements of the market considering the fabric technology a company can compromise onto. Later, speed grade, temperature, throughput, latency, etc. are well optimized to meet the market challenges. Once the specifications are fairly acclaimed by relevant teams of chip developing company, the architecture has to be planned accordingly. Presently Reduced Instruction Set Computer (RISC) architectures are being used and are available freely for the developers. Earlier CISC architectures were incorporated with the designs which is now replaced with RISC along with decision of floating points required for a processor. As the overall architecture of a single chip will be comparatively very large from a single piece of RTL code, because a processor has to perform many operations thus it equally carries those many numbers of RTL codes[13–20]. Briefly, a single RTL design of a processor is sub-divided into many other RTL codes called as micro architectural blocks. Each such micro architectural block can also carry an intellectual property (IP) available freely/cost effectively in the market. A behavioural representation of desired design is explicitly made to capture the actual essence of the design functionality.

2.4.2 Digital Logic Design

Once the architecture and specifications are well rehearsed by all the designers Fig. 4, now the next instance is to write up the specified hardware into code. These codes are nothing but the collection of keywords and user defined logic using hardware description language (HDL). User defined logic consists of the required digital logic based on Boolean expressions realized. RTL is expressed using with available standard HDLs or open source-built languages. VHDL and Verilog are the two HDLs used widely in the current generation. VHDL stands for Very High-Speed Integrated Circuit-Hardware Description Language. These two languages are also used for simulation and verification as well. Expressing RTL codes have different levels of abstraction. From top level of abstraction behavioral description is the most followed way of RTL descriptions, since this is the most compact way of writing the HDL. Accordingly, data flow level of abstraction describes how the data is produced from one end of the design till the output. Data flow is constructed mainly by the operators defined in the language library. Gate level or structural level description is the least way of describing code, here the HDL codes are written as per the Boolean equation which equivalently forms the logic circuit with some amount of logic gates. Gate level constitutes with huge number of temporary wires in the code which constitutes net delay for data to flow till the output, which is not desirable. Apart from the standard HDLs available, industry is gifted to have Bluespec or Bluespec System Verilog (BSV) and Chisel as an HDL way of describing hardware. BSV and regular Verilog has a huge difference when we consider the language data types, keywords, length of code written, etc. BSV is built so as to provide designer to carry out verification of their design exhaustively with lesser amount of time required to build the code rather than regular system Verilog or universal verification methodology (UVM). Other ways of generating HDLs are by using high level synthesis (HLS) where the hardware is basically described by C/C++ program and other is by using the feature of Simulink modelling provided by MathWorks MATLAB tool. Drawing design on object level and choosing a feature of generate HDL will convert the block level design to the specified HDL code. As we have indulged into industry 4.0, the recent ways of writing HDLs has the involvement of python programming language as well and known to be the finest way of writing any designer desired based digital circuit implementations. Coco-tb is one of python-based HDL testbench creating language which consists of concurrent based code execution while system Verilog and UVM are the most common languages used for verification, which we will be discussing more in brief in the coming sections.

2.4.3 RTL Verification – Micro Architecture or IP

After complete description of required digital hardware logic using HDLs the next passive but most crucial step is to check its behavioural exhaustiveness. Here we check the design by providing input vectors covering all the possible combinations, this helps to verify the design for its operative correctness whether it is responding with output as per the designed functionality for a particular input. As discussed in the above section system Verilog or Universal Verification methodology (UVM) are the two most used verification languages in the industry. These languages fall under the category of Hardware Design – Verification Language (HDL/HVL), apart from these we have open source developed language platforms such as BSV, cocotb, etc. while the objective for these remain the same i.e., to verify any RTL code

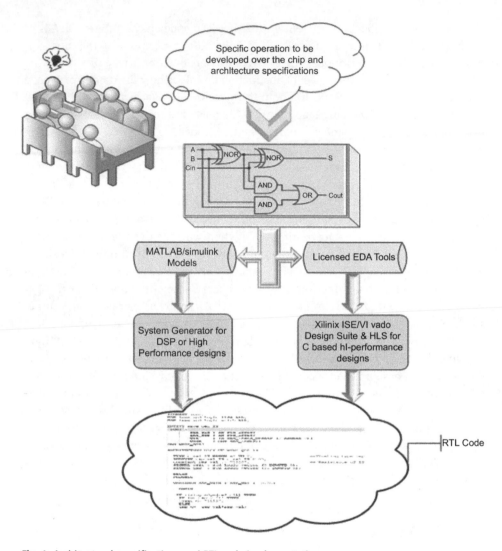

Fig. 4: Architectural specifications and RTL code implementation.

for its functional correctness considering all the corners of the design. The specific verification language has its own architecture for situating the RTL design into it. This verification code is basically known to be design under test (DUT), Fig. 5. Verifying the module (RTL) which has no other dependent sub module is basically an RTL verification, while several architectures/modules involved for a single design is provided for a verification team is known to be micro architectural verification which directly constitutes be the process of verifying an intellectual property (IP) i.e., the design is of some unknown company having its full rights over that design or property for their own perusal.

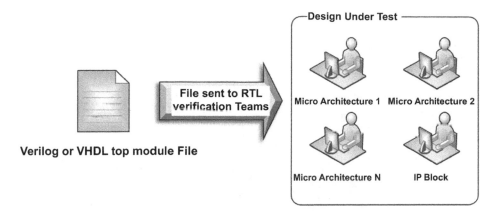

Fig. 5: Pictorial presentation of verification environment for existing top module.

2.4.4 Logic Synthesis and Netlist Generation

In this step the design is carried out for generating a netlist file. Netlist file consists of all the gates used by the design. To synthesize any design into gate level netlist file, the design has to undergo with three steps, which are translate, optimize and technology map. As we know that HDL languages have their own libraries defined as to be standard, similarly to synthesize any design there are standard leaf cell libraries provided by foundries which define size, area, cell type of a logic gate, etc. in their own library. These libraries are later used with the specific EDA and the fabric technology chosen by the user. Each fabric technology has its own standard cell library. This cell library during synthesis allows the tool to extract the logic and selects the specific logic gate required to convert it from RTL written code to actual gate (still in file format, not yet physically). Once the translation is completed, optimizing the design in best in class with respect to actual Boolean equation generated by written RTL code towards better optimized Boolean equation. Basically, it performs product of sums (PoS) or sum of products (SoP) methods. Now, when the logic has been optimized by considering cell library which is said to be fabric technology independent, is now mapped to fabric technology dependent. At this instance the tool tries to check if there are any user defined constraints, timing specification, port constraints, etc. and creates the Netlist considering these constraints with respect to cell library, Fig. 6.

2.4.5 Implementation, DRC Checks and Bit File Generation

Before the design is conjoined with the fabric industry for the initiation of chip sign off methods, we have an interesting and a crucial step where we try to implement

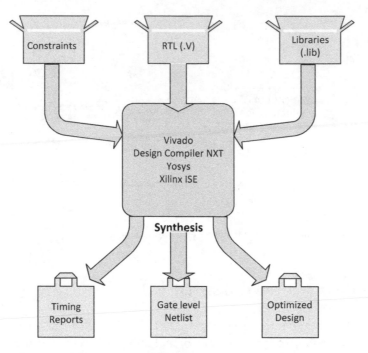

Fig. 6: Basic flow of a synthesis engine.

the design over FPGAs for prototyping and debugging signals using available tools. For this we initially start with design rule checks with the synthesized design which gives the result in percentile intending that the written design is following how much of the standard/company's rule so that it should not hinder with the succeeding steps. Now we chose the implement step from the tool which internally has three automated steps, translating available netlist and pin/timing constraints file into native generic database (NGD) file which constitutes of target device understandable design logic primitives. Followingly, map is the second step which tends to match the available ".ngd" file logic components with the available components on the FPGA or any other target, these components are basically I/O blocks, flip flops, LUTs, BRAMS, DSP tiles, etc. providing us with native circuit description (NCD) file. Finally, once the logic components are mapped it's time to connect these blocks called as routing. Constraints are well considered during routing steps which is the only step that tries to effect timing mostly in majority cases. Once the implementation step is fully covered with no errors, we are ready to generate the bitstream (.bit) file, is the only file that can be uploaded or program the target device (mostly FPGAs). Bitstream file has the all the necessary information required to configure the target device which was created by implementation tools, Fig. 7.

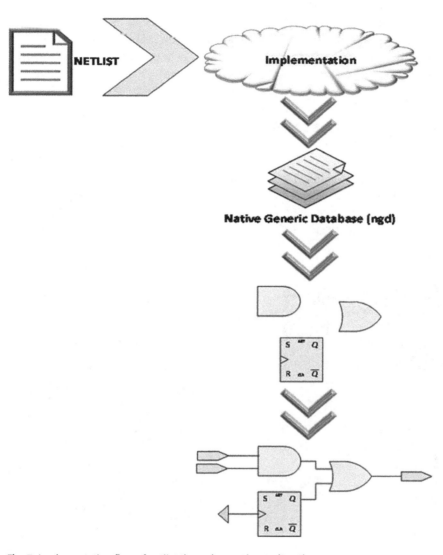

Fig. 7: Implementation flow of netlist through mapping and routing.

2.4.6 Simulation, Debugging Using Debug Cores and Transfer of Bit File to Back-End Teams

As the design is ready with the front end covered extensive files including bitstream, the design can now be verified by simulating it with the existence bit file or by creating a test bench for the top module RTL file by manually giving the input vectors. Other way is to capture the live signals from the targeted device by creating the debug core in the design. This debug core consists of all the internal nets from the required modules or architectures from the design project which we need to see in the debug waveform. The debug signals can be controlled by the additive feature provided by Xilinx Vivado Integrated Logic Analyzer (ILA) called as trigger control or capture control logic, where the user can specify the actual logic required to capture the signal at a particular event of other dependent signal in the capture window of ILA. Though adding debug will alter the design synthesis and constraints, re synthesizing and looping back to implementation step will provide a fresh bitstream file through which the target device can be programmed and can be performed the debug operations. Apart from Xilinx Vivado debug can also be carried over ChipScope Pro coming with Xilinx ISE, Fig. 8. As the market have been upgraded with latest trends and technology it is better to utilize Vivado for future designs as it has some advance add-ons which ChipScope pro doesn't come up with. It is to be noted that the debug process is only applicable for FPGAs and its development kit available.

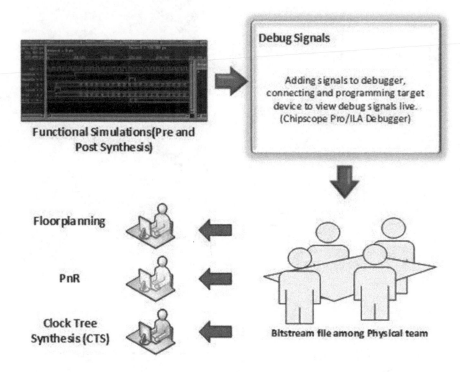

Fig. 8: Simulation and involvement of back-end teams.

2.4.7 Design Towards GDSII for Sign-Off

Till the previous section we have seen the generation of the bitstream file was the extent of the front-end flow of any semiconductor industry, from this section we can have a view on how the physical design teams are also involved to complete the entire production of a chip and generation of GDSII file, GDSII is an acronym for Graphical Design System (GDS). This file is the de facto standard for the industry to share the data related to layout of specific integrated circuit/chip. Considering the back end flow, the foremost step is the floor planning in which the designer calculates the aspect ratio of the chip die and core sizes so that implementing I/O pads, hard and firm macro cells, as well as power and ground pins. The major initial inputs for floor planning consist of RTL design netlist, constraint file, fabric technology file, foundry IPs and definitely a physical library file. Succeeding, we have come to the point where we place all the hard and soft macro cells with abutted and non-abutted combinations as per logic cell design requirement. Since there are cluster of open-source tools available in the market, one such tool named Open ROAD is a tool used for floor planning and almost entire physical flow. The next step after cell placement is clock tree synthesis (CTS) because timing is the crucial part for front end as well as for the back end VLSI. As bigger designs have more and more sequential elements that require synchronization, CTS is that step which provides clocking to all modules evenly by reducing latency and skews, if any. Now as the timing is newly implemented again in physical design, it will require new routing, as CTS will change the layout with respect timing routes for each cell. Routing needs to take care of used metal layers in the fabric technology and the entire physical flow has to be followed as per the industry rules/constraints which yields in the best production of an optimized chip. Basically, extraction of parasitic which consists of capacitance or resistances values perform major role for routing of wires. After the design has been fully routed, performing static timing analysis (STA) will provide us the total negative slacks if still exists in the design, if not the flow can be proceeded by viewing the entire physical layout by a layout viewing tools and perform spice simulations. This simulation helps to predict the performance of the complex designs. Although this simulation is just for observance of the analogous behavior of the physically implemented design. Finally, the routed netlist design is ready to be sent for fabrication, this netlist file is nothing but GDSII file format, which has the information in binary format. With this information the fabrication is ready to produce a chip integrated with implemented RTL designs.

2.5 Synthesized Modelling Approach for Custom Chip

Every man jack expects their product to be error free without any glitches and jittery free features, so does the RTL coding methodology. There are few whereabouts in the RTL development phase when the code doesn't seem to be built with bugs, but there are instances where the designer has made a huge blunder mistake which the design tool couldn't throw it as an error. As from the previous sections we have got to know about the synthesis, it is also important to know how does this part effects greatly to yield a good bug free design, which could later does not cause any fatal or false mismatches by the production or fab industries. As we write the HDL codes and move towards synthesis phase, pre synthesis and post synthesis are the two counterparts where the designer can compare the simulations from the both sides of synthesis. Any RTL design without any syntax error has an elaborated design which shows exact schematic as per coding abstraction, this is called as pre synthesis schematic. After completion of synthesis procedure, we are available with the netlist, basically this version of schematic available is known as post synthesis schematics. Since complex and bigger designs tends to develop billions of gates during synthesis, it is hard to sight the all-possible combinations of logic gates. Following few regulations could skip these kind of poor design techniques. Starting from the sensitivity list with the signals that are not used in an always block is not preferred, this could make the pre synthesis simulations run slower as the always block rarely executed than it is mostly necessary. Synthesis tool reads the signals present in the sensitivity lists and interprets it as a combinational or sequential or a latch based on the style of coding.

2.5.1 Three Cases of Sensitivity List

An always block with full sensitivity list containing all the signals/ports used within always block does not create a difference between the pre and post synthesis but when if one signal is missing from the sensitivity list that is being used in the always block makes the synthesis tool to perform exactly as per the code intention during post synthesis simulation but in pre synthesis the tool does not care if there are any changes over the signal that has been missed to be considered in sensitivity list, i.e. the value will not be observed for the missing signal. Apart from these if there is an always block with an empty sensitivity list which makes synthesis tool to implement that one such blocks for infinite loop during pre-synthesis simulations whereas post synthesis will again be as per the code intention. Finally, there is other important case when there is a mis-ordering of assignments of temporary signals/variables. If the temporary variable is used in the right-hand side without any value assigned to it, the tool infers it as a latch and holds the value for it with the last always block execution. In fig. we can observe the above three mentioned cases in RTL code in Fig. 9.

2.5.2 Usage of Functions and Case Statements

The most significant way of coding style is by defining the logic once in the function block and reusing the function block wherever applicable in the code. As the function block is synthesizable, it is acceptable unless the function block behaves as a combinational logic but when designers have to specify the sequential logic while

```
1    module example1 (a,b,c);
2      input a,b;
3      output reg c;
4
5      always @ (a,b)
6      begin
7        c <= a and b;
8      end
9    endmodule
10
11
12   module example2 (a,b,c);
13     input a,b;
14     output reg c;
15
16     always @ (b)
17     begin
18       c <= a and b;
19     end
20   endmodule
21
22
23   module example3 (a,b,c);
24     input a,b;
25     output reg c;
26
27     always
28     begin
29       c <= a and b;
30     end
31   endmodule
```

Fig. 9: RTL examples for sensitivity lists.

the function block is still to be a combinational, the tool will infer the block as a latch which is a bad way of coding and could mislead the output values due to metastability during simulation. This will again mismatch the values during pre-synthesis ad post synthesis simulations. There is no issue unless the function coded behaves as a combinational logic, but the same logic is again specified in the main code would infer a latch which is dangerous. Using case statements is most common while implementing FSMs (Finite State Machines) for hardware description. Every tool vendor has their own identification for directives which could confront the synthesis tool to perform the pre- and post-synthesis as per the code intention, such as for a full case and parallel case statements there are some unused states which can be assigned to don't cares, for this in Xilinx design suites using "//synthesis full case" will tell the synthesis tool to assign don't cares for the unused states. Similarly, for parallel case where all cases are treated parallelly i.e., the parallel case block will be inferred as a priority encoder even if their similar cases declared, but still the pre- and post-synthesis results are different. These similar issues can be observed while using the casex and casez statements.

2.5.3 Time Delays

Identifying the design flaws before the post synthesis simulation will make the designer understand what the checkpoints are for causing the mismatches. Still there are few coding methodologies where designers tend to assign timing delay during value assignment within the procedural blocks. In this case, once the value of the signal has changed its value which is kept in sensitivity list, its subsequent changes of value will not cause the execution of the procedural block unless the specified time delay has been completed. Such as, if there is total 25ns f delay for an assignment, the subsequent values of sensitivity listed signal will not enter the procedural block until 25ns has completed. This might cause the pre- and post-synthesis simulations to differ which is basically ignoring the events of sensitivity list signal. Though timing delays are not synthesizable, it must strongly to be avoided in RTL coding styles. Fig. 10, else the accurate model of RTL is never satisfied.

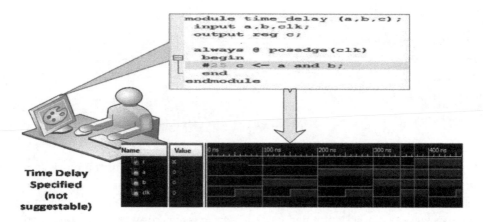

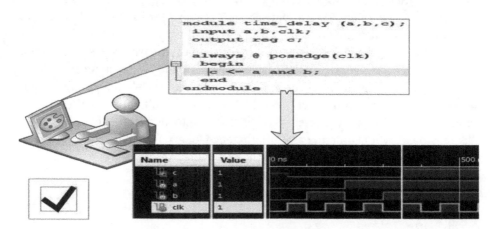

Fig. 10: Differentiated waveforms when time delay is specified.

2.6 Challenges in Industry 4.0

Since we have been canvassing about the emergence of RTL from semiconductor field methodically, it is equivalently important to know the involvement of semiconductor engineering into Industry 4.0. As the industrial revolution is redefining the marketing and manufacturing processes in a wider vision, if considered in what it is beneficial to an industry which is already sophisticated in digital realm. As the innovative approaches day by day is really required to make a great change inwardly, of an organization/industry, should equally merge the real and virtual world coherently. As the origin for these "faster" production methods tend towards only one term called automation. As the automation available today brings the new automation for tomorrow could only deal with the newer ideas for the re-outlining of industrial phases. Creating models and making business quickness, customizing the available prototypes with the required functionality and making it cost effective to the outer world, product quality and production efficiency or throughput is the epicenter of a present industry. As mentioned before, machine communicating with machines is the core work of the Industry 4.0. For this a big chunk of data has to be embedded into cores for processing the exact information related to the outside physical worlds and further connecting with the relevant equipment.

2.6.1 Advanced Data Analysis With Semiconductors

Total data being stored or managed by a chip is growing significantly year by year. Though the size of data is still small when compared to cloud data yet the chip has to have the ability to manage such huge amount of data or information to be processed. Since billions or trillions of gates will be used in a single chip to make a usable SoC or a specific IC with respect to its analog circuitry or RC networks which makes designer a challenging job to meet the constraints with a specific foundry. Extraction of exact data is crucial when consider bigger designs. Suppose a specific fabric technology consumes some x amount of voltage and the ideal cut off voltage is let's say 15% of x, now the designing team must follow this ideal voltage else the minimum power required for chip will get higher and higher which leads to more time consumption in optimizing these issues which may cause signal bouncing or metastability in core level. This also succeeds the issue from chip level to packaging industry and so on. For this industrial internet of things (IIoT) designs are being implemented which has enormous amount of data which could automate almost everything from vehicles to healthcare equipment and as well as drones, to name a few of them. This requires high power and highly efficient devices to start the production in which big data concepts plays a perfect role which shifts the designing team from solving repetitive issues towards meeting design specifications more easily.

2.6.2 Cluster of Larger Data Sets and Memory

Machine Learning (ML) is a big boost to the present generation technology because of its advent features which covers the maximum number of applications within a single category of industry, Fig. 11. As approaching towards bigger chunks of data for high-speed processing for a single silicon chip which can operate for several features such as voice recognition, speech recognition, facial identification, auto recommendations for mobile applications, health monitoring, etc. ML algorithms are most used for devising larger data sets into higher dimensional data sets (array), where data is sorted and separated accordingly. Higher dimensional sets or arrays require advanced mathematical calculations for data transformation like linear algebras are well pipelined by matrix operations. As we are considering enormous sized spaces of data, it directly portrays that more and more memory is definitely required for a robust operation supportively with big data. An extra memory component is compulsorily required beside the chip core for more speed. Larger caches are the next option where most of the designers opt out which gives internal computer memory for faster interconnects which is directly proportional to speed of processor. Because during boot up or refresh operation memory is not accessed and therefore cache is used mostly, this can allow processor to concentrate on its specific operation or algorithm which is primarily accessed by user. Apart from these there is an update in the design of integrated chip, where 3D integrated circuits are being proposed by companies which accelerates the sensing speeds and data acquisition. These 3D chips basically comes up with the higher cache density and required sensors forming a multilayered IC, in turn providing a bright and challenging future for semiconductor industries.

2.7 Applications of Industry 4.0

2.7.1 Various Industry 4.0 Applications

The fourth revolution in the Industrial sector is named Industry 4.0, which overcomes the challenges associated with manufacturing issues and improves the overall performance. Industry 4.0 brought automated manufacturing process, Machine learning, and Artificial Intelligence [19] taking care of quality control, and QR codes improving the logistic. As Industry 4.0 futuristically moves into the various domain of our everyday lives, many new emergences of applications that are quite difficult to predict today are being developed. As we enter the third decade of the 21st century, Industry 4.0 will establish a better future. Therefore, we can say that there is a major impact of Industry 4.0 on Industrial sectors

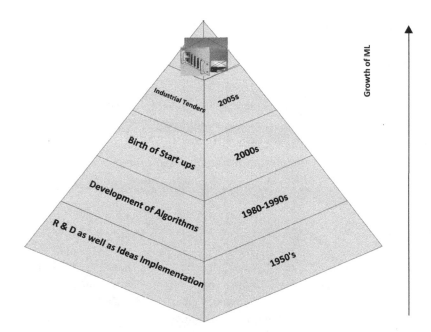

Fig. 11: Emergence of Machine Learning from past decades.

Cloud Computing

Today world economy is driven by the data and cloud computing plays an important role in that. From the mid-2000 use of cloud computing gets increased. Microsoft and Google followed Amazon's lead, as did several others such as IBM, HP, and Oracle. Cloud computing is following Amazon's early pay-as-you-use formula, which makes cloud computing financially attractive to SMEs (small to medium enterprises), as the costs of running a data center and dedicated infrastructure both IT and networks can be crippling. Many Industry works on real-time data, that will be store on the cloud for further processing. The consumer has to pay for the resource utilization. Cloud companies serve either Infrastructure as a Service or Platform as a service or system as a service depending on the requirement of the customer. Cloud computing technology further gives a ready platform to store this data and make it freely available to systems surrounding it.

Machine Learning

In the manufacturing industry, humans and machines have had a hands-on relationship. Machine learning technology [1] allows Industry 4.0 to gain a strong position in businesses and on factory floors. Machine learning comes from artificial intelligence that allows systems and algorithms to improve their performance automatically based on experience. The impact and involvement of machine learning and deep learning are rapidly growing in all sectors. Industry 4.0 also adopted machine learning algorithms to improve production. Cyber-physical systems, the Internet of things, and cloud computing are the current trends of Industry 4.0. Industry 4.0 is also called a "smart factory" where products will have the ability to collect and transmit data; communicate with equipment, and make intelligent routing decisions without the need for operator intervention. Machine learning provides improved visibility, statistics, and the ability to communicate problems in real-time. Machine learning reduces both processing time and human errors.

Neural Network

The cyber-physical systems (CPS) distinguish between microprocessor-based embedded systems and more complex information processing systems that integrate with their environment [2]. Robotics is an obvious example of a CPS. Robots can communicate with other machines as they are connected to the internet. Robots are good at sensing objects, gripping and transporting the objects. The robot can perform complex computations. In earlier days standalone robots were used in the Industry. They perform only like pick and drop or lifting jobs. Due to the involvement of the neural-network now Robots are capable to take the decision. As per the recent trends of industrial development use of robotics is rapidly increasing. Robotic automation is involved in product manufacturing to embedding NFC or RFID tags on the product and bring that product for logistic purposes.

Industrial IoT

In general, IOT systems are a network of physical devices like sensors. Most commonly used in buildings and cars. IoT mainly deals with big data, predictive analytics, and cloud computing. Its motto is to revolutionize digital services using framework structures, platforms and connectivity architectures. Industrial IoT is a subset of the Internet of Things [2], numerous sensors, Radio Frequency Identification (RFID) tags, software, and electronics are integrated with industrial machines and systems to collect real-time data about their condition, performance and status. Industrial IoT (IIoT) serves the different units and tasks associated with Industry. IIoT plays an important role in communicating important information in a better way and also to analyze and capture data in real-time. IIoT is capable to handle critical equipment & devices connected over a network. It has more sensitive and precise sensors. Smart health care systems are a good example of IIOT.

2.7.2 Semiconductor Industry-Specific Applications

Artificial Intelligence and machine learning are becoming popular due to the avail ability of high-speed hardware. Due to advancements in the semiconductor industry high-speed Microprocessors, GPU and SoC are getting used in several real-time applications like face recognition, speech processing, and many more complex and critical applications. Modern production control systems in workshops, such as fabrication units, are a perfect example of Industry 4.0, as it has become a synonym for increasing productivity in the 21st century by applying digital technologies in manufacturing.

Over decades Semiconductor Industry runs as per Moore's law, which states that the number of structures per IC doubles every twelve months. After few decades, Moore's law has slowed down cycles for new technology nodes, and adding more components in the IC causing more cost. Working on 10 nm technology and below can be possible due to revolution and automation in the semiconductor industry with minimum cost. In the semiconductor, industry efficiency of the production is increased because of the availability of wafer Production planning, digitization, and a high level of computerized automation. The overall production in the semiconductor industry gets improved by the smart production model like Cyber-physical Production Systems (CPPS) is adopted by the semiconductor industry. CPPS improves the business in terms of production to dispatch the material, enable cost-effective production of customized products, lower overall production costs, enhance product quality, and increase production efficiency. Now it is possible to add intelligence to materials and products with Industry 4.0 notions. Devices can communicate with each other, which increases flexibility and productivity at a much lower cost. Some devices are capable to hold their information without the need for additional electronics. The information about the cost of items gathered from the decentralized model and analytical software used in Industry 4.0 resulting in better intelligence for business strategy and product pricing.

2.8 Simulation Practices

The history of simulation can be emphasized as a drastic change in the strategies to verify the functional model of the design. The previous generation computers were used to verify the functionality of the current generation computer designs. In early 1950s there were no EDA tool introduced which was so advanced that it can built its own test cases or clock generation as required. Designers used breadboards and connect transistors and bunch of wires which used to tangle in the middle of simulations, but now the trend has made some easier way for designers where the larger designs can be simulated for its functional correctness by multiple means of simulation strategies. It is to be noted that design verification is entirely different than RTL simulations. Manually connecting analogue components and analyzing the design over breadboards, the current generation is available with the faster methods or computer aided ad hoc methods to solve any such problem while the ultimate goal of simulation practices is to omit the design errors prior to fabrication industry involvement in the production phase.

2.8.1 Linting and Its Importance

Lint or linting is a procedure of checking common and advanced erroneous lines on code. Linting has become a de facto term in industry for a software-based design verification which reflects good bug free code prior to simulation. This procedure can also be taken at the logic or gate level but as described, linting is popular for HDLs like Verilog, VHDL and System Verilog. As day by day the RTL codes are becoming more heavily complex with high level coding methodologies linting tools comes up with its own rules and guidelines that extracts bugs in code for better coding practices. Basically, linting is known to be a static verification technology. Past few decades linting was available only for checking syntax mistakes but succeeding as industry is moving into more and more automation. Linting tools have incorporated with the formal verification methodologies which has capability of verifying FSM based codes which has relatively more possibility of extracting errors like deadlock, metastability and stale states which are usually non executable states. One of the most noticeable features of lint tool is that it needs no test vectors like Sytemverilog requires, also unlike more coverage for logic code. If more errors are seen to be coming after linting a design, the design has to be re-debugged and generate newer synthesized netlist post the correction in RTL code lines wherever necessary, Fig. 12.

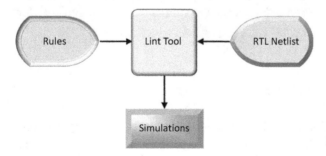

Fig. 12: Flowchart of linting procedure.

2.8.2 Logic or Gate Level Synthesis (GLS)

Now coming to actual simulation scenario, most of the companies prefer to build a testbench code to perform the simulation and verify the functionality of the design. The most common type of simulation metaphorically aims to watch the behaviour of the functionality of reset, clock and preset signals if present in the design. If there is any unintentional behaviour of these signals then it can be easily observed on the waveforms from the simulation tool. Also, if at all there are any DFT or scan chain structures present in the design post synthesis it can also be verified through GLS step. GLS can be considered much more complex method of verifying any design as gate delays, routing interconnects comes into picture which may cause setup/hold failures more frequently, due to which may cause a big confidence booster for any verification team that approves the design to be moved further for development post the dynamic circuit verifications i.e. GLS. By this we can wind up that GLS is mostly performed due to reset/preset performance verification, don't care conditional verification and timing verification mostly on asynchronous multi-cycle paths. Timing verification is included in this because in STA because asynchronous interfaces cannot be identified unless constraints are provided, for this we need a file called Standard Delay Format (SDF) which is prescribed to be verify timing issued in GLS phase. Mainly this phase targets the maximum operating frequency of design by which asynchronous paths are compared and alternatively synchronizer flip flops are constructed by CDC tools and propagation of don't care values to be verified is the main core of the GLS. Don't care propagation may cause due to timing issues, uninitialized memory units and latch inferred logic from RTL code. Verifying all this design can be taped out for the further process in design or R&D processes that comes right before sign off phases.

2.9 Conclusion

This chapter shows a revelation of the present and future of Industry 4.0 specifically to semiconductor industry explaining the industrial steps in order to design custom RTL for the chip. Additionally, this chapter helps the researchers to understand the importance of semiconductor industry in respect of Industry 4.0. Wherein, all efforts also show a new path towards the chip design flow involving custom RTL design, simulation, synthesis and timing analysis of a particular module and its uses in semiconductor engineering. A better clarity on the various challenges and applications of industry 4.0 has also been portrayed with custom RTL design and their ASIC/FPGA implementation incorporating all the functionality like lesser area, higher speed and low power consumption.

References

[1]   Ghaffari, A., & Savaria, Y. (2020). CNN2Gate: An Implementation of Convolutional Neural Networks Inference on FPGAs with Automated Design Space Exploration. Electronics, 9(12), 2200.

[2]   Gilchrist, A. (2016). Industry 4.0: the industrial internet of things. Springer. https://doi.org/10.1007/978-1-4842-2047-4

[3]   Mahapatra, A., & Schafer, B. C. (2019). Optimizing RTL to C Abstraction Methodologies to Improve HLS Design Space Exploration. 2019 IEEE International Symposium on Circuits and Systems (ISCAS), 1–5.

[4]   Nosalska, K., & Mazurek, G. (2019). Marketing principles for Industry 4.0 – a conceptual framework. Engineering Management in Production and Services, 11 (3).

[5]   Karacay, G. (2018). Talent development for Industry 4.0. In Industry 4.0: Managing the digital transformation (pp. 123–136). Springer.

[6]   Y. Lu. (2017). Industry 4.0: A survey on technologies, applications and open research issues. Journal of Industrial Information Integration, vol. 6, (pp. 1–10).doi: 10.1016/j.jii.2017.04.005.

[7]   Lu, Y. (2016). Industrial integration: A literature review. Journal of Industrial Integration and Management, 1(02), 1650007

[8]   Yu, B., Xu, X., Roy, S., Lin, Y., Ou, J., & Pan, D. Z. (2016). Design for manufacturability and reliability in extreme-scaling VLSI. Science China Information Sciences, 59(6) (pp. 1–23).

[9]   Ma, Y., Suda, N., Cao, Y., Seo, J., & Vrudhula, S. (2016). Scalable and modularized RTL compilation of convolutional neural networks onto FPGA. 2016 26th International Conference on Field Programmable Logic and Applications (FPL). (pp. 1–8).

[10]   Zhang, C., Li, P., Sun, G., Guan, Y., Xiao, B., & Cong, J. (2015). Optimizing fpga-based accelerator design for deep convolutional neural networks. Proceedings of the 2015 ACM/SIGDA International Symposium on Field-Programmable Gate Arrays. (pp. 161–170).

[11]   Chu, S. L., & Lo, M. J. (2013). A New Design Methodology for Composing Complex Digital Systems.

[12]   G. De Micheli, G. (2009). An outlook on design technologies for future integrated systems. IEEE Transactions on Computer-Aided Design of Integrated Circuits and Systems, 28(6) (pp. 777–790).

[13]   Bening, L., & Foster, H. (2001). RTL Methodology Basics. Principles of Verifiable RTL Design: A Functional Coding Style Supporting Verification Processes in Verilog. (pp. 43–68).

[14]   Da Silva, K. R. G., Melcher, E. U. K., Maia, I., & Cunha, H. do N. (2005). A methodology aimed at better integration of functional verification and RTL design. Design Automation for Embedded Systems, 10(4).(pp. 285–298).

[15]   Shmerko, V. P. (2004). Malyugin's theorems: A new concept in logical control, VLSI design, and data structures for new technologies. Automation and Remote Control, 65(6). (pp. 893–912).

[16]   Takeda, E., Ikuzaki, K., Katto, H., Ohji, Y., Hinode, K., Hamada, A., Sakuta, T., Funabiki, T., & Sasaki, T. (1995). VLSI reliability challenges: From device physics to wafer scale systems. Microelectronics Reliability, 35(3). (pp. 325–363).

[17]   Hirose, M. (1988). Future very-large-scale integration technology. Materials Science and Engineering: B, 1(3–4). (pp. 213–220).

[18]   Goser, K., Hilleringmann, U., Rueckert, U., & Schumacher, K. (1989). Rough overview of VLSI systems, basic circuits, and technologies also offers some concrete examples of realized integrated circuits. IEEE (Institute of Electrical and Electronics Engineers) Micro;(USA), 9(6).

[19]   Rabbat, G. (1988). VLSI and AI are getting closer. IEEE Circuits and Devices Magazine, 4(1), (pp. 15–18).

[20] D. Mills and C. E. Cummings. (2005). RTL Coding Styles That Yield Simulation and Synthesis Mismatches. Princ. Verif. RTL Des. (pp. 43–68).doi: 10.1007/0-306-47631-2_4.

[21] https://www.designnews.com/electronics test/what role-will-semi-play-future-industry-40 (Accessed on 06/05/2021)

[22] https://sst.semiconductor-digest.com/2016/12/industry-4-0-what-does-it-mean-to-the-semiconductor-industry/ (Accessed on 06/05/2021)

[23] Panda, S. K., & Panda, D. C. (2018). Developing high-performance AVM based VLSIcomputing systems: a study. In Progress in Computing, Analytics and Networking (pp. 315–323). Springer.

Ram Singh, Rohit Bansal, Vinay Pal Singh

Industry 4.0: Driving the Digital Transformation in Banking Sector

Abstract: Industry 4.0 or Digital Revolution is changing the way we live, changing associations with customers and organizations, which unavoidably infer that both existing business techniques and monetary administrations are not excluded from this change. In the monetary world, accelerated digitalization has made banks genuinely re-examine standard strategies, which suggests that they need to respond quickly and gainfully to the solicitations of their clients while offering secured and direct organizations for use. Security and trust are at this point key determinants, and banks have made innovative monetary organizations and things over the span of ongoing years, including got structures that constantly guarantee the data and money of clients. Regardless, really like any cutting edge uprising, both Industry 4.0 and its impact on the difference in the monetary region pass on with themselves both positive and unfriendly after effects of this change. Digitalization of the monetary territory is in as far as possible, with the way that this connection also consolidates other fragment segments of Industry 4.0, for instance, blockchain networks, man-made cognizance, IoT, biometrics, cooperation of deals with a record with FinTech associations, plan of the stage, and various organizations for the Generation Z and other. In this paper, we investigate how expanded rivalry, new enactment, and every one of the progressions that accompanied digitalization, will influence the financial area in the impending time frame, will the financial area look fundamentally changed in the forthcoming years, and will, in spite of every single innovative change, human factor, trust, security actually be the key determinants.

Keywords: IR 4.0, IoT, big data, digital revolution, GAFA, digitalization, chatbots, FinTech

3.1 Introduction

3.1.1 Digital Driving Forces

The previous decade has seen a far and wide acknowledgment of computerized innovations. As we enter the Fourth Industrial Revolution, organizations across businesses

Ram Singh, School of Business, Quantum University Roorkee, e-mail: ramsinghcommerce@gmail.com
Rohit Bansal, Department of Management, Vaish Engineering College Rohtak,
e-mail: rohitbansal.mba@gamil.com
Vinay Pal Singh, School of Business, Quantum University Roorkee

https://doi.org/10.1515/9783110725490-003

and verticals are going to AI, ML, IoT, Big Data, and other present day advancements to smooth out their tasks. With regards to India, the public authority itself has taken a large group of drives to advance the huge scope selection of innovation in different areas. The rush of advanced interruption has influenced a few areas like assembling, retail, coordinations, friendliness, travel, and even instruction [3]. Almost 39 percent of organizations in India intend to put 8 percent of their yearly incomes in advanced projects by 2021 and the nation would before long observer a mushrooming of really computerized undertakings. Despite this advancement, banking is one area that actually remains generally immaculate by innovation. While a few public and private area banks are digitizing their administrations, we have only start to expose what digitization can accomplish. It turns out to be very obvious when contrasted with the financial scene around the world. Worldwide banks are changing their plans of action to offer an advanced first encounter to clients. Additionally, buyers are getting to know banks that are conceived advanced. Taiwan's O-Bank, for instance, has no physical branches and offers its types of assistance through an application and a 24x7 video contact focus. The expanding dependence on innovation is an obvious sign that the eventual fate of banking virtual [2]. The financial area in India has encountered an extreme change lately and the actual meaning of banking has changed. Before, net banking was restricted to sending or accepting cash. Commonly, the cycle a few days, however today, we have made a lot of headway in this space through IR 4.0, be it making new records or applying for individual credits, advanced banks have empowered shoppers to profit banking administrations through a gadget that is strategically located in the palm of their hands. Some open area banks are in any event, taking this computerized change to an unheard of level, the public area bank is looking at the disposal of actual check cards, and there will be a restricted need to have plastic cards in the following 5 years.

Furthermore, it's not simply the banks, NFBCs and little loan specialists are likewise taking the advanced course to furnish clients with an issue free, consistent experience. In the previous three centuries human progress has gone through three modern transformations, while as per the assessment of the world's driving financial experts, the fourth mechanical upheaval is continuous or as it is likewise famously called the Industry 4.0. Every one of these mechanical upsets was described by mechanical advancements that keyly affected the improvement of whole humanity [5]. What is normal for Industry 4.0 is that it effectively in various manners influences all business exercises, while at the same time creating advanced and different advances, yet in addition influencing the whole way of life on the planet. After the cycle of globalization and the association of the world into one worldwide market, which brought about an unhampered development of business, another period started, which can be known as the time of advanced change. The essential trait of the new, computerized age is that it takes new measurements and new structures, starting with one day then onto the next. In spite of the fact that, by their design, a type of business, and different attributes, banks are more averse to acknowledge changes, yet they have to a great extent changed their business to changes in the business climate and, thus, received and applied certain cycles forced by the digitalization interaction. The consistent cycle of making new financial items and administrations that are straightforwardly connected to the digitalization interaction is an obvious indicator that the financial area has genuinely perceived the forthcoming changes, which positively bring about the making of upper hand and a superior position of the market. Obviously, huge rivalry available, the rise of different administrations offered in corresponding with banking administrations; show market over-burden, and surely, it presents the greatest test for banks in the impending time frame. It must be particularly borne as a main priority that in the present circumstance, banks are rivalry to each other, yet additionally cutting edge organizations managing comparative administrations that have arisen over the most recent couple of years and began offering this sort of administration are likewise an opposition to them. These organizations have their own installment frameworks and client data sets, which brings about an allowance of part of the banks' benefit. This plainly implies that banks need to work seriously on advancements in the financial field and foster new business techniques and models that will be adjusted to new requests available.

Positively, notwithstanding the exercises identified with the presentation of new administrations, just as their acclimation to the market, the spotlight should in any case be on the customer of the bank for example banks should focus on the ideal nature of administrations that will fulfill existing customers and furthermore pull in new customers to the bank. It is vital that the change cycle of banking administrations is joined by consistent tuning in to the market and customers' necessities since it ought not be failed to remember that all banks have and will have customary clients later on, who will absolutely utilize the standard financial administrations, just as customers who surely won't utilize advanced financial administrations in a specific timeframe. On account of the entirety of this, the banks should look for the ideal measure in changing their specialty units and presenting computerized bank offices that would totally supplant HR. Basically, the cycle of digitalization in the financial area is, other than the incredible benefits for banks and their customers, additionally carrying with it certain difficulties that banks need to deal with. The advanced change will additionally diminish the banks' benefit in the forthcoming years, which will be a result of considerably more noteworthy rivalry and the continuation of the decrease in banks' edges. Banks, in the entirety of this, are attempting to rival the presentation of inventive administrations, accessible through cell phones, notwithstanding, in this, a ton of their incomes are taken by little advanced showcasing organizations that are progressively engaged with work that was saved distinctly for business banks until yesterday. In the event that banks intend to situate themselves available in a sufficient manner and adjust to new changes, it will be important to rapidly change the plans of action they work on, by changing themselves from only monetary organizations into foundations whose stage will be founded on information examination and offer of proper items and administrations with which they will contend available, just as by opening up more noteworthy freedoms for participation with FinTech organizations [8].

3.1.2 The Development of Fintech in India

Notwithstanding this, India has likewise seen the rise of various FinTech NFBCs. As indicated by industry gauges, our nation has arisen as the world's second-biggest Fin-Tech center point, following just the US. Among different elements like rising Smart-phone proprietorship and more profound web infiltration, government support has assumed a crucial part in driving the development of the Indian FinTech area. Organizations with a yearly turnover of Rs 50 crore should utilize government-supported advanced installment stages like Aadhaar Pay, NEFT, BHIM, and so on, and no additional charges will be forced on either clients or dealers. Be that as it may, in a country where money keeps on being the essential method of exchange, government hardware alone can't assist the country with accomplishing its monetary consideration objectives. This is the place where B2B organizations come into the image and can impel enormous scope selection of advanced installments in the country [6].

3.1.3 Virtual Banks Image

Technology is ready to carry a significant change to the financial area. All things considered, man-made reasoning and different AI-sponsored applications have effectively begun changing the operational perspectives and client assistance angles [7]. According to a Juniper Research report, upwards of 2 billion Indians will utilize computerized banking before the current years over. Notwithstanding the difficulties like security dangers, doubt towards advanced banking, absence of monetary education and mindfulness, developing proof consequently uncovers that computerized banking is turning into the most favoured type of banking even in India. There are various advances accessible these days that can help during the time spent the computerized change of the financial field. A decent computerized change banking technique ought to incorporate applicable innovations that can bring the most incentive for both the bank and its clients. Among the moving innovation arrangements, we can recognize:

3.1.3.1 Computerized Reasoning

AI in banking is addressed by chatbots or online collaborators that assistance clients with their issues by giving vital data or executing various exchanges. Aside from this, AI can be utilized with the end goal of information examination and security. For example, recognize illegal tax avoidance by examining client information inside a few seconds.

3.1.3.2 Blockchain

The execution of blockchain in banking can bring about a superior interface, more precision, and got information and exchanges. Furthermore, blockchain arrangements will make exchanges and various activities straightforward, working with coordinated effort. There will be no requirement for the intermediation of the outsiders, subsequently raising the degree of trust from the clients. It can likewise impact cloud advancements and move to decentralized ones, which will bring about higher security of the information and assets.

3.1.3.3 Internet of Things

IoT is useful with ongoing information examination, in this way makes the client experience more close to home, and banks can give exclusively customized offers. Also, on account of wearables, clients can without much of a stretch and flawlessly make contactless installments. Aside from this, IoT is useful with hazard the board and admittance to stages; the verification cycle can be upheld by biometric sensors that make access safer and ensured.

3.1.3.4 Distributed Computing

Cloud processing is another innovation that can assist saves money with getting proficient and allows them an opportunity to give more advancements, just as have better efficiency, improve tasks, and quickly convey items and administrations. Distributed computing, too as IoT, can assist with hazard the executives and establish a protected climate for the clients and inward bank frameworks. With everything taken into account, the computerized change of the financial field will bring extraordinary advancements that will change the picture of banks we know these days. Advances bring various freedoms both for banks and their clients by getting individual information, expanding straightforwardness, and allowing an opportunity to oversee reserves whenever and anyplace.

The financial digitalization measure unavoidably carries with itself new components and opportunities for extending new financial administrations, and in accordance with that, new chances for expanding the vital exhibitions of the business, particularly the benefit of the bank. One of the essential chances for improving business results is reflected in the innovative opportunities for better communication with the bank's customers, just as a more definite and exact knowledge into their requirements, propensities, and potential outcomes. Advanced change generally infers a method of business activities that identifies with the difference in inward and outside techniques. With the help of these methodologies, the utilization of current advances will give more proficient work to representatives and surprisingly better quality relations with customers [9]. A fundamental change will possibly occur if all assets are utilized to execute new methodologies. Banks of things to come will positively be proficient, present day, and mechanically outfitted bank offices with no long lines and administrations will be offered through suitable self-support machines and PCs that will zero in on customers, which is the primary objective of the bank. The fundamental objectives that banks need to carry out later on through the interaction of computerized change are to adjust the administrations and methods of overhauling with better approaches for business activities, the presentation of administrations dependent on close to home insight, just as the use of the idea of moral banking.

3.1.4 Digital Bank & the Use of Artificial Intelligence

The key change concerning the computerized bank offices is in cooperations for example in the connection between the bank and customers. Today it is empowered through different tablets and PCs where all financial administrations can be performed, which can likewise fill in as promoting loads up simultaneously. Also, another watchword for an advanced bank office, notwithstanding collaborations, is the expression of development for example a consistent cycle of development that other than administrations offers different exercises in the advanced bank office. Different administrations may incorporate a piece of bank office for understanding papers and free utilization of the Internet, show of nearby displays, collaboration with insurance agencies, and so on The advanced bank office of things to come should offer to its customers an alluring and present day banking climate dependent on connections and development that will offer customers proactive and fast admittance to the data they need [11]. The eventual fate of banking is unquestionably in advanced advances, yet just the change from customary banking to computerized banking will be empowered, obviously, the customer and the bank connection will keep on being in the core interest. Also, computerized bank offices will absolutely be founded on new techniques for correspondence, individual access, simpler admittance to data and administrations, and appealing and intuitive branches. Banks that will regard these standards will definitely enjoy a similar benefit and will unquestionably have more prominent consumer loyalty, natural business development, and acknowledgment on the financial market. Quite possibly the most regularly referenced subjects recently is the job and the idea of man-made reasoning, and it is notable that the information we have are the motor these days and that man-made brainpower is equivalent word with the new force of current occasions. As well as improving client manuals and giving fast data, without mistake, computerized reasoning can likewise be utilized to robotize measures at the bank. Basically, the digitalization cycle is travelling toward the path where all that can be digitalized can't avoid being digitalized. Nonetheless, we must not fail to remember that the key thing is believe that can't be digitalized so all that can't be digitalized will turn out to be vital and required, which are sure feelings, innovativeness, creative mind, morals, compassion, instinct, and trust so calculations can supplant everything with the exception of what depends on this.

With regards to banks, a few banks have effectively made "Visit Banking" stages dependent on computerized reasoning, through which customers speak with the bank through Facebook and Viber applications. "Chatbots" are the initial phases in the utilization of man-made brainpower in the financial area. Their job is in that banks rapidly and productively react to clients' requests about items and administrations. Man-made brainpower will absolutely be huge both in the space of adjusting the client and in the space of counselling the client. "Psychological Computing", which is a blend of mechanical technology, man-made reasoning, and work with a lot of information, is seriously utilized today on the planet [15]. There is additionally a stage being ready for the utilization of psychological automatization that is further developed than the advanced mechanics of the cycle and will actually want to work with unstructured information. The key is that the joining of man-made brainpower brings added development, speed, and dexterity to monetary activities, and yet it keeps up trust as the essential hypothesize of banking tasks. The objective of man-made consciousness is that its application gives positive ramifications to the current brand and notoriety that the organization has available, to expand business effectiveness, yet such that it is coordinated and applied to monotonous exercises, and become standard, so representatives can be centered around more inventive work that brings added esteem both exclusively and for the organization where they work.

3.1.4.1 GAFA (Google, Amazon, Facebook, and Apple)

Later on address the greatest rivalry for banks in the piece of giving installment administrations. The inquiry emerges whether, by applying the new PSD 2 Directive, these organizations will remove a portion of the income from banks and how much they will assume control over their business. As per certain studies, it is normal that these organizations can partake in the profit from banks. Above all else, the reconsidered PSD 2 Directive on Payment Services empowers outsiders to offer monetary types of assistance, regardless of whether they are fintech organizations or monstrous like Google, Apple, Facebook, and Amazon. GAFA (Google, Apple, Facebook, and Amazon) are as of now attempting to make their own portable installment framework, inside their foundation. Instagram, alongside Facebook, is attempting to introduce installments inside its foundation, while Amazon is searching for an accomplice bank, and Google has TEZ India and the new Google Pay Wallet in the United Kingdom and the United States [11]. The truth of the matter is that four banks previously joined the drive that UBS and IBM organizations dispatched in 2016 to assemble another world exchange stage dependent on blockchain innovation.

The new stage, called "Batavia," was made to give open admittance to associations of all sizes anyplace on the planet, and can uphold exchange money exchanges across all methods of exchange, regardless of whether merchandise are moved via air, land, or ocean. Stage's Batavia will probably dispose of the need for taking care of and contrasting archives, permitting purchasers, merchants, and their banks to execute exchanges with a significant degree of effectiveness and straightforwardness. Remembering that blockchain advancements require its cooperation organizations in its turn of events and execution, it is totally expected that they participate in the "value-based banking" income, and considering the reality the blockchain innovation attributes that are, a blockchain PC document comprising of information blocks which are interconnected, it is consistent that it is improvement and application can't sidestep huge IT organizations, and, in accordance with that, remove a portion of their benefits.

3.1.5 Blockchain Technology & Cryptocurrencies

3.1.5.1 Blockchain Technology

One of the patterns that are normal for the digitalization of banking is unquestionably the presence of blockchain innovations and it is identified with the presence of cryptographic forms of money. The presence of computerized cash for example cryptographic forms of money is an inescapable outcome of the Industry 4.0 unrest and the advanced economy. A great deal of cryptographic forms of money showed up, with just some of them encountering prominence and full certification. Perhaps the most acclaimed cryptographic forms of money are surely Bitcoin, which has been mainstream over the most recent couple of years. The fundamental trait of this kind of cash is that it capacities totally freely from financial specialists and from the measure of customary cash available for use. Verifiably, Bitcoin has acquired its full confirmation over the most recent couple of years, basically because of the development of its worth. Bitcoin was made as a prize for members who dealt with certain social ventures and got grants as digital currencies for their work. Relations among members and their commitment to the framework work on the "shared/P2P" guideline) [17]. Blockchain can affect business similarly that the Internet has impacted correspondences in light of the fact that these organizations are changing totally the way how things are done and they are essential for the computerized advances that change all areas. Various banks got included and joined particular organizations occupied with these organizations, and in this path stages for computerized financing of exchange dependent on blockchain innovation are together evolved. Furthermore, such foundations empower, other than banks, additionally different organizations, and particular establishments to have less difficult financing and exchanging measures, accordingly decreasing the danger of inability to satisfy commitments. The benefit is that they can arrange and satisfy orders, arrange the terms of economic accords, and access monetary administrations offered on the stage with complete security and trust.

3.1.5.2 Use of Blockchain Technology

The circumstance on the world's market is to such an extent that organizations are as yet trying different things with blockchain, so they have not yet moved it to finish creation. Blockchain innovation will change banking by and large, since it empowers the speedy, secure, and modest exchange of quick shipments, without mediators, and it nearly bars the chance of hacking. Blockchain will change banking, similarly as the Internet has changed correspondences and the media in light of the fact that blockchain permits everybody to send cash very quickly and with moderately low expenses. Numerous banks have effectively begun presenting blockchain innovation in certain spaces of their business. A few enormous national banks are among them, like the Central Banks of Russia, the Netherlands, and Canada. After the underlying undertakings, the end was made that the new innovation, other than its benefits, has certain deficiencies. The referenced banks have brought up the issue of whether the blockchain can react to their requirements and can it truly expand effectiveness and decrease costs? It worked out that innovation is progressing nicely, yet there are still weaknesses that are average for creating advances. One of the principle protests identifies with the speed of exchanges when they are executed in genuine volume for example in ordinary business. A huge component of the blockchain that got a transformation its application is the way that the requirement for any trust between the gatherings associated with the exchange is abrogated. It isn't even important to confide in an outsider that ensures the dependability of exchanges. Rather than trust, the job was taken over by cryptography. The SWIFT framework presently being used for monetary exchanges is as yet predominant and it figures out how to do an enormous number of exchanges with high dependability, however one ought to have as a primary concern that SWIFT is really the result of a similar privately owned business.

3.1.5.3 Advantages & Disadvantages of Digital Transformation

Computerized change of the financial foundations brings a great deal of new freedoms for the clients, singular ones, little organizations, and immense companies, just as banks themselves. Sharp procedure and consistent enhancements can bring about various benefits:

– **Comfort**
Personal and friends ledgers are accessible on any gadget, the solitary things you need are an Internet association and a couple of taps on the screen. This brings

more consumer loyalty as they can continually stay focused of their record adjust and deal with the data on their own profile (add new postage information, messages, phone numbers, and so forth) Moreover, there is no compelling reason to go to the bank to get checks as they can be in a split second shipped off your email address.

– Every minute of every day Service
Online financial administrations are accessible day in and day out lasting through the year, even at the ends of the week. There is no compelling reason to remain in lines and trust that the bank will open to direct certain activities. It's an enormous benefit that accompanies advanced arrangements.

– Efficient
Another benefit, you save a ton of time as you approach the record from home. It is very helpful as already you could squander a little while at the bank to simplify activities, and now it is done consistently from home or whatever other spot where there is an Internet association.

– Robotized Transfers
Direct banks can give limitless mechanized exchanges (acknowledge finance stores or give programmed charge installment) with no extra expenses for the administrations even to outside monetary establishments.

3.1.6 Simpler Management

Online records can be handily overseen, in spite of the fact that they require more data than conventional banks. Clients can add data themselves or straightforwardly contact online partners to offer help on the recent concern. Also, payee data is held inside the framework, there is no compelling reason to return information for the accompanying installments, and so forth albeit the benefits are noteworthy, and they work with the work by and large, there are still a few weaknesses that follow the interaction. Here are the primary ones:

3.1.6.1 Security Issues

Cybersecurity is quite possibly the main issues that organizations and establishments are attempting to survive. Indeed, even modern programming that secures speculative information can't totally shield accounts from con artists, phishing, programmer assaults, and so forth.

3.1.6.2 Administrations

Nowadays, not everything banks can offer a wide scope of online administrations. In any case, there are some that require your quality at the customary banks.

3.1.6.3 Exchanges

Complex exchanges may likewise require the presence at the bank office. Additionally, worldwide exchanges are impractical with all advanced banks. The quantity of hindrances is somewhat low, and it won't be long until they vanish. Benefits assume control over inconveniences and make every day activities a lot simpler, decrease costs, save client's time, and figure out how to offer types of assistance proficiently.

3.1.7 Challenges of Digital Transformation in Banking Sector

The eventual fate of advanced financial change is noteworthy, and it is anticipated to completely change the picture of customary banks, just as carry more administrations to the clients. Yet, today banks face certain difficulties that are hard to survive:

3.1.8 Rivalry with Non-Financial Institutions

Amazon is attempting to give banking freedoms to its clients. Facebook permits clients to make moves straightforwardly to others' ledgers, subsequently avoiding keeps money with regard to the cycle. Be that as it may, banks are more directed foundations, in this manner safer. It is significant for them to go advanced faster.

3.1.9 Online Payments

Not each bank can uphold on the web/portable or contactless installments. The justification this is that a bank can't give another alternative however actual installment as it doesn't have a safe online stage, assets, and talented group to make it conceivable. It gets considerably harder to rival accessible online administrations like ApplePay and PayPal.

3.1.10 Advances

The framework that gives web based financial administrations ought to be continually refreshed and has an effective security level. Network protection is quite possibly the main issues these days that banks should mull over as a matter of first importance. Once more, to beat this test, the bank needs to set up an itemized technique, pick suitable advancements and track down an expert group that can undoubtedly transform thoughts into reality with the most recent programming and accessible instruments. The interaction of computerized change is fairly interminable as innovations improve; the frameworks ought to be continually refreshed. New advances will allow an opportunity for new administrations to give the idea that again will require consistent overhaul and backing. Computerized change in banking will accelerate in the forthcoming years. Monetary foundations should consider the present status of undertakings, work out a significant methodology, and utilize the correct devices to prevail later on. The change venture is certainly not something simple and includes a great deal of assets, however thus, the administrations could be more gotten and conveyed faster, and in particular, upgrade the degree of consumer loyalty [14]. As advances create and upgrade, the financial business will likewise improve and better step by step.

3.1.11 Conclusion

Investigating the latest things in the advancement of digitalization of banking administrations, it is obvious that banks should change their plans of action and adjust them to either speed up market changes or to shape a coalition with enormous innovation organizations, just as with more modest organizations that have reciprocal arrangements actually like the banks. Moreover, banks should act proactively towards administrative specialists and lessen their working expenses so they can play a market game. From an authoritative perspective, all future changes will be founded on the innovations and abilities of banks to rapidly beat new techniques for handling continually expanding measures of information. Numerous banks become accomplices with FinTech organizations and have joint interests in mechanical tasks [9]. Underline that corresponding to the digitalization interaction, it is important to satisfactorily manage the monetary structure to wipe out or moderate deliberate dangers. As a matter of first importance, it is important to secure customers and their information in the computerized economy, to fit organizations that arrangement with comparative exchanges, and apply similar guidelines to all market members. This suggests that it is important to track down the suitable harmony between rivalry, advancement, security, and customer insurance.

It is deliberately significant that the effect of Industry 4.0 on the change of the financial area should not risk security to the detriment of rivalry and development. From all the previously mentioned, plainly Industry 4.0 impacts the change of the financial area, with the way that specific imperatives and the legitimate system that would forestall any conceivable adverse results of this cycle should be dealt with. Banking of things to come will unquestionably be set apart by a further interaction of digitalization of banking items and administrations, which implies that banks that need to take part in a market game should put resources into new advancements, which will surely mean extra incomes they can rely on, yet in addition the expenses of extra ventures at this phase of the bank's change. If we need it, get it or not, new advancements are relentlessly coming, bringing totally new ideas that expect us to change the manner in which we think. In the event that we figure a similar route as we did as such far and we anticipate that the new technology should bring us just advantages, without clear investigation and contribution in all cycles, all things considered, we are on some unacceptable way. Digitalization of monetary administrations isn't just about the utilization of new innovations yet it additionally suggests a totally unique methodology and a totally new idea of deduction for all members. Having as a top priority the sped up transforms, we can say that the future financial will most likely appear to be unique in the forthcoming years, it is sure that computerized reasoning will supplant various cycles, add to the speed increase of correspondence and banking administrations with better calibre [13]. Eventually, we can say that other than the sped up digitalization measures, both in banking and in different zones, the human factor will keep on assuming a vital part later on and the attention will stay on trust, security, and customer of the bank.

References

[1] Anouze, A.L.M., & Alamro, A.S. (2019). Factors affecting intention to use e-banking in Jordan. *International Journal of Bank Marketing*, *38*(1), 86–112

[2] Asha, N., Lahari, N.M., & Amrutha, S.N. (2020). Renovating Indian banking and financial sector through Industry 4.0 powered technologies. *Journal of Business and Management*, *22*(4), 8–12.

[3] Balcerzak, A.P. (2016). The technological potential of the European economy. The proposition of measurement with the application of multiple criteria decision analysis. *Montenegrin Journal of Economics*, *12*(2), 1–11.

[4] Bilan, Y. (2019). The influence of industry 4.0 on financial services: determinants of alternative finance development. *Polish Journal of management studies*, *19*(1), 70–92.

[5] Buer, S. V., Strandhagen, J. O., Chan, F. T. S. (2018). The link between Industry 4.0 and lean manufacturing: Mapping current research and establishing a research agenda. International Journal of Production Research, *56*(8), 2924–2940.

[6] Capgemini (2012). The digital advantage: How digital leaders outperform their peers in every industry. https://www.capgemini.com/wpcontent/uploads/2017/07/The_Digital_Advantage__How_Digital_Leaders_Outperform_their_Peers_in_Every_Industry.pdf (Accessed on 06/ 05/2021)

[7] Eleni, D. (2019). The impact of AI in the banking sector & how AI used in 2020. *International Journal of E-Business Research*, *15*(4), 24–39.

[8] Fettermann, D. C., Cavalcante, C. G. S., Almeida, T. D. De, Tortorella, G. L. (2018). How does Industry 4.0 contribute to operations management? Journal of Industrial and Production Engineering, *35*(4), 255–268.

[9]   Hair, J., Anderson, R., Tatham, R., & Black, W. (2010). *Multivariate Data Analysis with Readings*. US: Prentice-Hall: Upper Saddle River, NJ, USA.

[10] Hanjun, S.S.C. (2018). An empirical analysis of a maturity model to assess information system success: A bank-level perspective. *Journal of Behaviour & Information Technology*, *2*(1), 5–12.

[11] Igaz, A.T., & Ali, A. (2013). Measuring banks service attitude: An approach to employee and customer acuities. *Journal of Business and Management*, *7*(2), 60–66.

[12] Islam, N., & Borak, M.A. (2011). Measuring service quality of banks: An empirical study. *Research Journal of Finance and Accounting*, *2*(4), 74–85.

[13] Josef Horak. (2016). Does industry 4.0 influence efficiency of financial management of a company. The 10th International Days of Statistics & Economics, Prague, September, 8–10, 2016.

[14] Lu, Y. (2017). Industry 4.0: A survey on technologies, applications, and open research issues. *Journal of industrial information integration*, *6*(1), 1–10.

[15] Mackay, N. (2019). The impact of industry 4.0 on the transformation of the banking sector. *Research Journal of Finance and Accounting*, *4*(1), 11–24.

[16] Mayank, P. (2019). Digital transformation in finance. *Journal of Research in Finance*, *2*(3), 28–36.

[17] Schwab, K. (2018). The Global Competitiveness Reports, Chapter 3: Benchmarking Competitiveness in the Fourth Industrial Revolution, 2018. Geneva: World Economic Forum. Retrieved from http://reports.weforum.org/global-competitiveness-report-2018/chapter-3-benchmarking-competitivenessin-the-fourth-industrial-revolution-introducing-the-global-competitiveness-index-4-0/ (Accessed on 06/ 05/2021)

[18] The Banker. Branches of the future-changing for the digital age. May, 2018. Page 21.

[19] The Banker. Publication from the Financial Times. May, 2018. pages 20–25.

[20] Xu, M., David, J. M., & Kim, S. H. (2018). The fourth industrial revolution: opportunities and challenges. *International Journal of Financial Research*, *9*(2),1–6.

Sahana. P. Shankar, Deepak Varadam, Harshit Agrawal,
Dr. Naresh E

Blockchain for IoT and Big Data Applications: A Comprehensive Survey on Security Issues

Abstract: Today's world is full of new technologies and advancements in every field. With more unique, advanced technologies coming up daily in the modern world, along with the ease of living comes narrowing of the privacy of an individual. With more technologies coming up, taking more information about an individual, the privacy of the individual is get violated and, thus, brings a threat to the security of the individual too. In this chapter, two such modern technologies that have now become an inevitable part of the human life are discussed along with what security measures they lack, why or how they lack these security measures, and what can be done to secure these. But before going to their security aspects, let's first know something about the technologies like what role do these play in society and what is their importance in daily human lives.

Keywords: blockchain, big data, IoT, IoHT, ICT, IFCIoT

4.1 Introduction

4.1.1 Internet of Things (IoT)

The idea of the Internet of Things (IoT) was started being discussed as early as 1982, gaining the interest of more and more researchers, and thus evolving slowly. It finally gained a boost somewhere between 2008 and 2009 (as estimated by Cisco) [1]. Internet of Things (IoT) is a collection of interrelated objects such as computers, machines, etc., including humans, where there is an exchange of information on the network between two devices or a device and a person. The collection and distribution of information are mainly done by embedded systems such as sensors and processors, continuously collecting data from the environment. The IoT is a beneficial and necessary technology these days; not only it helps to make the daily lives more

Sahana. P. Shankar, Dept. of Computer Science and Engineering, Ramaiah University of Applied Sciences, Bangalore, India, e-mail: sahanaprabhushankar@gmail.com
Deepak Varadam, Harshit Agrawal, Dept. of Computer Science and Engineering, Ramaiah University of Applied Sciences, Bangalore, India
Dr. Naresh E, Dept. of Information Science and Engineering, Ramaiah Institute of Technology, Bangalore, India

https://doi.org/10.1515/9783110725490-004

comfortable and smarter, but it also gives a significant input in the business sector with smooth transactions and making more intelligent business decisions.

4.1.2 Applications of Internet of Things

As discussed in the above section about what the Internet of Things is, it can be seen that IoT networks can be extensively used in many fields like agriculture, healthcare, home automation, data collection, weather analysis, etc. Even smartphone connections also come under the Internet of Things as a smartphone also contain various sensors which collect data. This data is processed by the smartphone's processor accordingly, and then it may or may not be sent to some other smartphone or some data server or any other node in this IoT network. Thus, it won't be incorrect to say that in a few years, the Internet of Things not only evolved significantly but also led humans towards a smarter and more comfortable (though not secured) life. Here are some important fields of application of the Internet of Things along with how and why IoT has become an essential and integral part of these fields:

4.1.2.1 Healthcare

Technology growth in healthcare has been very rapid in the past few years. Amongst those many technologies, IoT also plays an important role in healthcare nowadays. Sensors like heart rate sensor, blood-pressure sensor, pedometer, blood oxygen level sensor, and others are now integrated into the devices as small as smartphones and fitness bands, thus enabling people to track their fitness daily, suggesting them workouts, diet-plans and other activities to maintain their health.

4.1.2.2 Logistics

When it comes to logistics, the first thing that comes into mind is smart transportation. Though it is a vast field on its own, it is just one part of where IoT is used to automate things. IoT has enabled the automation of a standard appliance as small and as usual as a refrigerator used in the warehouses to store perishable goods to the automation of transportation vehicles (self-driving cars), automation of warehouses (smart-homes), and even automation of entire cities (smart-cities).

4.1.2.3 Agriculture

IoT nowadays has been extensively used in agriculture. As using various sensors like humidity sensor, sensors to detect a concentration of multiple components of soil i.e., phosphorous, nitrogen, etc. can be used to measure the nature of the soil calculatedly and can also predict the chemical requirements of the land with their required quantity to make the soil sufficiently fertile for the crop. IoT not only enables the farmers to know what nutrients to add to the soil but also the appropriate amount of those. As previously, due to the lack of knowledge of the proper amount of nutrients requirement of the land, farmers used to add extra nutrients, which in turn harmed both the soil and crop. Now farmers can use this technology and get to know about the health of their land, they don't anymore need to go to the soil test centers far from the villages.

4.1.3 Big Data

Data, in today's world, is the biggest and an essential resource to humankind and thus can also be of great danger if used with wrong intentions. "With big data comes big re-sponsibilities," large information, which is principally alluded to as the tremendous hurl of information that can't be overseen by conventional information taking care of strate-gies or methods. The field of large information has a significant part in different regions, for example, agribusiness, banking, information mining, training, science, account, dis-tributed computing, showcasing, medicinal services stocks. Large information examina-tion is the technique for taking a gander at huge information so as to discover shrouded designs, immense relationship, and other important information to use to manage this information and settle on basic choices with it. There is a huge growing enthusiasm for big data on account of its quick turn of events and various territories of uses.

4.1.4 Applications of Big Data

With the population all around the globe increasing vigorously day by day, the amount of information related, contained, produced, required, and utilized by an individual is also growing with even a higher rate. This increase in population can be practically imagined by the fact that previously data was measured in Giga-Bytes or rarely in Tera-Bytes (access to this much data/storage was limited to very few users only). But now, Tera-Bytes seems to be a regular storage unit, and even smart-phones have that much capacity. The world has moved to even larger unites like petabyte, exabyte, zettabyte, and even yottabyte. Some of the real-world applica-tions of big data are:

4.1.4.1 Government

Big data analysis and processing play a significant role in government sector func-tionalities. It helped in Barack Obama's re-election campaign in the USA. Also, by the processing and analysis of big data, the government can identify the emerging issues and needs of the public in the society and can center itself to work on them.

4.1.4.2 Social Media Analytics

One of the largest providers of big data is social media. By using social media analytics, it helps out many companies to know about market needs, trends, behaviors, and competitions. This further helps the companies to keep track of the feedback of people towards their products.

4.1.4.3 Banking

Bank database is another significant source of big data, with the information of lakhs of customers stored in it. Big data analysis in banking helps to know about the client's reaction about the bank's services and to maintain statistics of how many new customers are there, how many old customers discontinued the services, and it also helps in fraud detection.

4.1.5 Security Issues in the Internet of Things

There are many reasons like unsecured network, centralized architecture, no authentication, no encryption for the lack of security in IoT networks. But it's not like that the security measures aren't taken in IoT network for no reason. The fact here is the lack of resources in IoT nodes. IoT nodes are the end devices/sensors collecting and sending the data. Being small in size, they have limited resources in terms of both computation capabilities and storage capacities, while implementing authentication or encryption techniques require both computation capabilities and storage capacities. Even if the limited amount of resources available in IoT nodes, if authentication or encryption is implemented in IoT, it will result in the latency between data collected by the sensor and transmission to the user.

4.1.6 Security Issues in Big Data

Talking about security issues in big data, as discussed above, the size of big data is nowadays measured in zettabytes, the utilization of client information perpetually raises security issues. By revealing shrouded associations between apparently irrelevant bits of information, big data analysis might uncover delicate individual data. Research shows that 62% of brokers are mindful of their utilization of vast information because of protection issues. Further, re-appropriating of information examination exercises or conveyance of client information across divisions for the age of more extravagant bits of knowledge likewise enhances security dangers. For example, clients' profit, investment funds, home loans, and protection approaches wound up in inappropriate hands. Such occurrences fortify worries about information security and dishearten clients from sharing individual data in return for altered offers. Thus, security maintenance in big data is very crucial to ensure the privacy of the individuals.

4.1.7 Possible Security Solutions for Security Threats in IoT and Big Data

There area few many methodologies that can be used to secure IoT networks and big data. These include:

4.1.7.1 Network Security

Securing the network on which the IoT devices are connected, or the big data is transferred, will increase the reliability of data being sent over it making the whole IoT network and big data transmission secure.

4.1.7.2 Authentication

Another way is to devise the authentication management system on the network. So, only authorized users to have access to the data shared.

4.1.7.3 Encryption

All the data sent over the network can be encrypted using some basic cryptography techniques like RSA cryptography.

4.1.7.4 Interface Protection

Similar to that of authentication, an interface can be created to access the data which only the trusted users can access.

4.1.7.5 Blockchain

It is known to be the most secure architecture to date, blockchain was introduced as the underlying security architecture for the first cryptocurrency Bitcoin. Blockchain, in laymen's terms, is a digital ledger that keeps track of the transactions and their information. But, having a unique architecture, these transaction details are stored very securely.

4.1.8 Architecture of Blockchain

While taking the name into consideration, blockchain can be considered as a simple chain of significant blocks chained together in a particular manner, which is also considered as one of the greatest inventions after the internet. If we look deeper, it is a collection of different technologies that are already present and known around for a long time, which are- private key cryptography, A distributed network with a shared ledger, and a way of providing computation and secured storage for the transactions and records related to the network. The cryptographic network technology of the blockchain is used in the Bitcoin for trusted transactions between any two parties.

The blockchain is a distributed chain of blocks that store transactional information. It is an instance of information that is stored in each device. Due to this distributed architecture, if someone ever tries to manipulate the data of a block in the blockchain, he/she has to do something called Proof of Work (discussed in the next section). He/she has to not only do this for the instance of blockchain stored in his/her personal machine but for all the instances available in all the machines. Apart from this distributed architecture, one more object of interest in blockchain is the design of each block. The information stored in each block of blockchain can be divided into three major sections:

The first section contains the necessary transactional details such as date, time, place, amount, currency type, etc. of the transaction.

The second section contains the information of the individuals (or organizations) involved in the transaction. To maintain the security of user information, and the authenticity of the transaction, the digital signatures of these individuals (or organizations) are stored in the block.

The third section contains the information about the block itself, which includes its unique hash value and the hash value of the last added block.

4.1.8.1 Proof of Work and Adding a Block to the Blockchain

In order to add a block to the blockchain, the hash value of the block needs to be generated, which can only be generated by computing a complex mathematical function. The given mathematical function is so complex that its computation requires a lot of resources. This function takes the data to be stored in the block (first and the second section of information) as the input and then produce a unique hash value to be allotted to the block. The computation of this hash value is known as proof of work. So, if someone wants to add a block to the blockchain, first he/she should provide all the transactional details along with the digital signatures of involved individuals (or organizations). Proof of work has to be done to calculate the hash value, which is then stored in the block itself along with the hash value of the last added block, and finally, the block can be added to the blockchain.

4.1.8.2 Blockchain Security Mechanism

At first, as already mentioned, any block in the blockchain doesn't contain the user information directly, instead of stores a digital signature. But the main reason blockchain architecture is considered one of the most secured architecture lies in the idea of having proof of work concept and of storing the block's hash value and the hash value of the last added block. If any person tries to access and manipulate the transaction details of a block(already stored in the blockchain), first he/she has to do the proof of work to generate the new hash value. Even if the person succeeds in doing that, the current block's hash value now changes. Therefore, the older hash value, which was stored in the next block, doesn't match with this new hash value, and thus, the attack can be identified and nullified too. But if the attacker anyhow succeeds to successfully change the hash value stored in the next block, also by doing the proof of work again, the hash value of this next block will now be changed. Thus, the hash values of all the blocks stored in the blockchain need to be manipulated. If the attacker is so dedicated to even do that, since multiple instances of blockchain are stored on various devices, the attacker has to change the hash values in every instance of the blockchain from all the devices. Having such a comprehensive level of security, it may even take years for the attacker to manipulate a single transaction completely. Even the amount of money, time, and efforts he/she will be spending on the resources for this may not be even worth the transactional amount. Figure 1 below explains the security mechanism in blockchain technology.

Recent trends in popularity of Blockchain technology can be seen in the Fig. 2 below. In [2], authors have done an extensive survey and developed statistical data in the context of the popularity gained by blockchain technology in past few years.

The above graph shows the trends in the investments (around the globe), which have been made to implement blockchain technology in many fields and also in the research on how and what other fields can be facilitate by blockchain. It can be seen that every year the amount of investments is going up only and by 2019, the blockchain industry was expectedly worth around 400 million USDs.

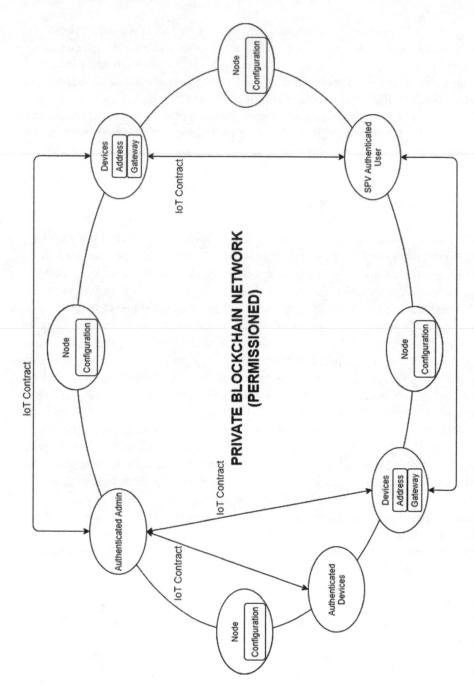

Fig. 1: Security Mechanism in Blockchain Technology.

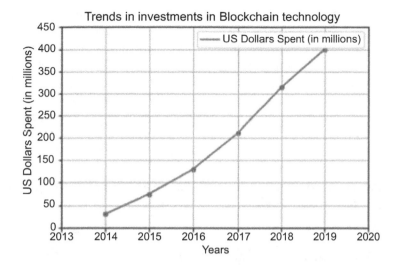

Fig. 2: Trends in investments in Blockchain technology.

4.2 Application of Blockchain in IoT and Big Data for Healthcare

4.2.1 IoT in Healthcare

The importance and current trends in the application of IoT in the healthcare domain have been discussed in the previous sections of the chapter. Apart from that, IoT networks are constantly evolving day by day are getting more and more integrated into the healthcare domain. From the as basic functionality as counting the number of steps of a person using a pedometer, to measuring the heart rate or the blood oxygen levels can all be now done using small IoT nodes. Not only it helps the user/patient only to keep track of his/her fitness, but it also helps the doctors to make a proper diagnosis since more and more data is now available. Also, this data can be grouped, processed, and can then be used to train a machine learning model to perform the diagnosis at a very initial stage. Many such models do even exist today, which with more data, can make some advanced diagnosis too.

A new term has been recently coined for this integration of IoT and healthcare i.e., IoHT (Internet of Healthcare Things), in which various healthcare sensors, systems, and personnel are connected over a network. This network does the exchange of vast amounts of data from one node to another. Numerous human services associations are as of now utilizing IoHT, from checking new-borns to following stock and looking after resources. Mirroring this intrigue, a report by the board counseling firm Frost and Sullivan predicts that the quantity of IoHT gadgets will ascend from about 4.5 billion of every 2015 to up-wards of 30 billion by 2020 [3]. IoHT can assist the healthcare functionalities majorly in two ways – one in clinical services and others in support operations. In clinical services, IoHT is aiding in healthcare by providing essential services like Remote Patient Monitor-ing (RPM), using which the doctors can now monitor the patient from his/her home after being discharged. RPM not only helps the doctors to keep track of the symptoms of the patients but also allows them to suggest some preliminary procedures in case if anything is wrong. Along with that, it can also help to keep track of the patient's medicine dozes, as many times, patients aren't very regular with the medications given to them or take less than suggested doses. IoHT likewise helps clinical preliminaries by intently following fundamental signs and some other markers applicable to the examinations, for example, glucose levels, pulse, circulatory strain, weight patterns, and so forth.

4.2.2 Security Issues with IoT in Healthcare

Since the IoT network is collecting very crucial and private information about peo-ple, it is essential to keep this data safe and secure. As privacy and data, protection comes at the top in software usage concerns. Being an unsecured network, many companies hesitate to implement IoT networks in their healthcare devices, and this is the reason that IoT is not being used to its full potential in healthcare yet.

Various methods to implement security in IoT networks are already discussed in **1.1.7**. In [4, 5], authors have used the methods other than blockchain to secure the IoT networks which include, ensuring the security of system on which the data is transferred, encrypting the data collected, and restricting the access to the data to authorized users only, by implementing authentications. The target of the dis-cussed architectures is to provide privacy-protection by authenticating the users.

4.2.3 Blockchain for Addressing the Security Issues with IoT in Healthcare

Blockchain is already famous for its secure architectural features. Thus, it can be used to secure the IoT network in which the healthcare data is collected and trans-ferred on, and the privacy of the patients can be maintained. Blockchain-enabled IoT devices could use the feature of seamless data exchange provided by IoT net-works while also improving healthcare for all.

Besides only securing the patient's medical data, since the architecture of the blockchain is such that once a block is added to the blockchain, it's almost impossible to alter it. It also helps in maintaining the authenticity of the transaction (using smart contracts), and neither the supplier or receiver can make false claims about any of the commodities or services.

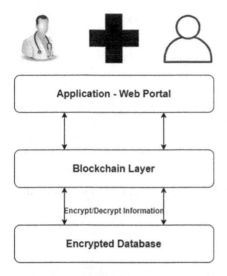

Fig. 3: Three-tier architecture using blockchain.

Figure 3, visually defines the three-tier architecture to improve security in healthcare fields using blockchain.

The main issues while implementing IoT in healthcare applications include the transfer of data in real-time, managing a distributed network, handling the heterogeneous nature of the data coming from various sensors, usage of data in diverse applications. One of the possible solutions to these issues came out to be cloud computing in the past few years. Still, cloud computing also had problems like latency in data transfer, difficult management of the devices, etc. Therefore, after the introduction of fog computing in 2014, people are moving towards fog computing for the implementation of IoT in various fields, including healthcare. With the computations being done on the nearby edge devices in fog computing instead of on a remote cloud server, the latency is reduced. Also, using fog computing, an interface for the authentication of the users who access the data can be developed, leading to a somewhat secured implementation [6, 7].

In [8], authors have proposed a recommender system, that recommends the type and availability of food in different areas based on (1) the diseases that the person is suffering from, (2) the diet plan the person follows in his/her daily life. The proposed system uses the pre-fed information in the system by the user and then uses IoT devices like heartrate monitor, blood pressure sensor, etc. to continually gather the data and improve the recommendations. When the person is travelling, the system also recommends the complimentary food available in that region, if the planned or regular food items aren't available in that region.

In [9], authors have proposed an autonomous wireless body area network which can be used in remote patient monitoring. To make the system independent, the sensors which are deployed on the various positions on the body of the patient, are solar-powered and Bluetooth low energy-based. Thus, the proposed system can be used for a long term monitoring of the patient with the almost negligible cost of energy consumption. The only essential requirement of the system is that the patient must go out for a short period.

In [10–12], the authors have described how the blockchain technology works and what are the security aspects of it. The information division and storage mechanism is also explained. Authors also discuss how a block is added to the blockchain, the concept of work of proof, and smart contracts. Thus, the core architecture, the idea of how blockchain is providing security and integrity to the data is explained. Further, this secure architecture is linked to IoT network to aid the healthcare domain. Before that, it is essential to know what are the basic requirements (in terms of security) of IoT in healthcare. The basic requirements are – security, interpretability, data sharing, and data access. Then the authors describe how blockchain can be implemented in IoT (using cloud or fog) and fulfil all the requirements mentioned above.

In the past few years, with the advancements in the technological industry, it also revolutionized the healthcare industry. The introduction of small health monitoring sensors is probably the field of most interest nowadays. It can be heart rate count, daily steps count or some other metric for the self fitness tracking, or it can also be a collection of crucial patient's health data for remote patient monitoring. But, the problem is a small size; these sensors are not able to provide security to the data collected and sent over the network. In [13, 14], authors have proposed an architecture to implement this remote monitoring system and integrate it with blockchain technology to provide security to the patient's health data and therefore, his/her privacy too.

4.2.4 Big Data in Healthcare

Big data consists of a massive amount of data that may be structured or unstructured. This data is so broad that it is almost impossible to process it using the traditional methods/techniques. In healthcare, a massive amount of information is collected every second. This data can be the patients' details who visit the doctor, the scan reports, or any other test reports done. But the amount of this data is gone to a new level with the introduction and boost in the usage of wearable devices collecting healthcare data. The exciting thing about these wearable devices is that they record the user's health data every second. When this data combines with the usual clinics' data, the amount of the total healthcare data rises to a new extent. In healthcare, information is being produced ceaselessly, every time a visit is paid to any essential consideration doctor or a pro, every excursion to an emergency room (ER) or any surgeries that is experienced. Apart from these, there are health insurance companies as well that also contribute to this data. Thereafter coming to the unstructured part of this collected data, since the structure used by two different clinics, or two wearable healthcare devices from two different manufacturers (or sometimes even from the same manufacturer too), or two different health insurance companies can be all different. Since no standard structure is yet defined to store the medical information due to the difference in the gathered data (fields) from different bodies. Therefore the collected data from various sources have a different structure, and thus, the data becomes unstructured.

Using various devices (including IoT) like mobile phones, smartwatches, health bands, sensors, etc. A considerable amount of data is getting collected every minute, and all this data since coming from varying sources becomes unstructured. Therefore, it is crucial to process this data and convert it into meaningful data. By doing this, machines can be trained to identify the symptoms of various diseases at an early stage only. Also, it would be beneficial for the researchers and medical professionals to gain meaningful insights from structured data rather than from an unstructured one. In [15–20], authors have discussed why big data analysis is essential in healthcare, how can it further revolutionize the whole industry and what are the different ways to implement this in healthcare. The authors also discuss the current trends in healthcare and big data.

4.2.5 Security Issues with Big Data in Healthcare

In the recent few years, there has been a significant hike in the number of data breaches, which also increases concern about the whole architecture of how this healthcare data is collected and stored. As a violation in the healthcare data of one can result not only in severe economical-losses and legal penalties but may also result in a loss in concern of health. Nowadays, whenever visiting a doctor, they already have most of the information about the patient. Also, they make the patient sign the document to acknowledge and understand their terms of the privacy policy. Apart from this, the integrity of the stored data must also be maintained i.e., the data once stored shouldn't be able to be tampered by anyone. But at times, there are situations where healthcare data of a patient must be visible to a specialist, to study and diagnose the patient more efficiently. Therefore, the data should be visible to the authorized users, and that too should be done in as little time as possible. Because in case of emergencies, delay of providing this information by even a second can cost huge health penalties for the patient. The patient records ought to be predictable and accessible across institutional limits, and access to this data ought to be constrained by the patient [21–22].

4.2.6 Blockchain Solutions for Security Issues with Big Data in Healthcare

The Blockchain innovation permits us to sort out information such that exchanges can be confirmed and recorded while getting agreement from all gatherings included. The innovation utilizes the idea of a legitimate record that monitors all occasions.

In [23], the authors have proposed a three-tier solution for securing the health-care data by implementing blockchain as the underlying architecture. In the proposed architecture, the first tier/layer is the application or web portal, which can be accessed by a patient, a doctor, a clinic, a health insurance company, or even the wearables collecting the healthcare information. This portal also allows providing a filter of information access based on user type (provides authorization). The second layer consists of the blockchain itself in which the data collected from the application or web portal provided by the patient or doctor or any other healthcare institute is put on the blockchain. (the procedure on how to put data in a blockchain is already explained in the previous sections). Thus, the integrity and security of data are ensured. But to allow the authorized users to access this data, in the third layer, the information is encrypted and stored in a database (cloud platforms too can be used for this). Whenever requested by the authorized users, the data is decrypted and, corresponding data in the block is displayed on the application or web portal itself.

"With big data comes big responsibilities". The distinction of Blockchain advancement and the gigantic level of its application results with much-progressing examination in different sensible and legitimate locales. Albeit still new and in the testing stage, the blockchain is being viewed as a progressive arrangement, tending to present-day innovation concerns like localization, trust, personality, information proprietorship and data-based choices. Simultaneously, the world is confronting an extension in amount and assorted variety of advanced information that is created by the two clients and machines. While effectively looking for an ideal approach to store, sort out and carry out Big Data processing, the Blockchain innovation can come in giving colossal information. Its proposed game plans about the decentralized organization of private data propelled property objectives, IoT correspondence and open associations' progressions are critically affecting how Big Data may progress. In [24–29], authors present the novel courses of action identified with a part of the Big Data districts that can be improved by the Blockchain advancement.

Stats and Evaluations

In [30–33], authors have done extensive surveys on the comparison of security in the IoT networks and big data. Surveys are conducted with the integration of blockchain and without the integration of blockchain. The Tab. 1. below highlights a few of the challenges, their explanation and potential blockchain solutions.

Tab. 1: Challenges solved by blockchain.

Challenge	Security without blockchain (encryption, etc.)	Security with blockchain
Costs and capacity	Since several IoT devices and data transferred over them has increased a lot. Therefore, it is required to increase the capacity of networks.	No requirement for a centralized entity: gadgets can convey safely, trade an incentive with each other, and execute activities naturally through shrewd agreements.
Deficient architecture	Due to a large number of devices in the network, each device's bottlenecks are combined. They are thus increasing the aggregate bottleneck of the system.	Secure informing between gadgets: the legitimacy of a gadget's character is confirmed, and exchanges are marked and checked cryptographically to guarantee that lone a message's creator could have sent it.
Cloud server downtime	Cloud servers are sometimes down or unavailable to provide services.	Since the records are on many devices, disfunctioning of a single device won't affect the system.
Manipulation	Since the data's integrity isn't confirmed, it can be manipulated easily.	Provides integrity to data by giving concepts of proof of work and smart contract.

4.3 Application of Blockchain in IoT and Big Data for Logistics

4.3.1 Introduction

The logistics industry is one of the biggest industries worldwide generating huge revenue for countries worldwide. There is an estimate that the logistics industry by the year 2023 will reach a value of 15.5 trillion dollars [6]. In the present day, there are many changes faced by the logistics industry due to increased costs, labor problems, and pressure from the government, the great influence of technology and automation that is being adopted by the fresh startups. This has led to the promotion of the logistics industry from the traditional sector to a digital one. The main driving force behind this happens to be artificial intelligence, machine learning, Internet of things. Many logistics and transportation industries have adopted Data Science and the Internet of Things among other competing technologies to revolutionize the way business is being done. The logistics industries have discovered that there is a huge wealth of data that is residing in the company over the years which isn't utilized well. Mining this hidden data generates a huge amount of knowledge, which could help the company in various ways. Many of the applications of Big Data and IoT are overlapping. In many cases, it is seen that these two technologies are jointly applied to achieve the best possible solution.

There is the various application of Data Science and IoT to logistics management. Some of the major ones include the following:-
– Optimization of the warehouses in logistics
– Total delivery time estimation
– Optimization of the delivery path leading to a reduction in freight costs
– Supply to demand dynamic price matching
– Demand forecasting
– Better maintenance by extension of the asset's shelf life by finding usage data patterns
– Management of the delivery of perishable goods
– Online location tracking and monitoring

Logistics management involves billions of shipments every day all across the globe, which includes data such as location, contents, weight, size, origin, and destination. The tracking that is happening is mere of the data. The value that can be extracted out of this data is still untouched completely. Big data can help logistics companies to increase their operation efficiency, customer experience, and also to develop new business models that are information oriented.

Operational efficiency is achieved in logistics management where the data is used to increase transparency levels, in the optimization of resource consumption and to improve the quality of process and performance. Customer experience can be enhanced by exploiting data to increase customer retention and customer loyalty, optimization of customer service and interaction, and perform accurate customer targeting and segmentation. Develop new business models by capitalizing on data by generating novel revenue streams from new data products and the expansion of existing revenue streams from already existing products. Let us now look into a few of these applications in greater detail.

4.3.1.1 Operational Efficiency

The planning of shifts optimally is one of the important tasks of logistics management. It can be planning, for example, shifts for the warehouse employees. Proper shift planning to meet customer demand has to be done carefully. Lower site profitability and more expenses have to be borne because of overstaffing. On the other hand, operating the warehouse with fewer employees leads to a negative impact on the employee as well as customer satisfaction. In either way, it is going to harm the business. In the case of data science, it is very important to make use of huge amounts of historical data collected over the years, to predict the demand at every point of sale reliably. The big data algorithms make use of various data factors such as opening and closing hours, the arrival time of new goods to the warehouse, holidays of the market, diversions on the roads, real-time traffic data, weather conditions. In this manner, predictive analysis algorithms can be employed to improve operational efficiency.

4.3.1.2 Enhancing Customer Experience

Big Data solutions focus on providing customer service based on individual customer behavior and needs. This is in contrast with the traditional way where one strategy fits all is being used. One of the popular examples of logistics management could be trying to mitigate the "out of stock" conditions to satisfy the customers. One of the most disappointing experience for a shopper is when he finds a perfect item that satisfies his needs in all dimensions but is not available in stock currently. With increasing competition in various sectors of the logistics industry, it is estimated that in a few cases there are only weeks between the first design of a product to its availability in the warehouse shelf. With the increase in the demand for online shopping, there is a boom in the logistics industry too. The profitable key factor for business here is to be able to correctly predict the demand. If the demand predicted is too much, it would be leading to locking potential capital, on the other hand, shortage of items would lead to the company losing the goodwill of customers.

4.3.2 Security Issues in Application of Big Data and IoT Solutions for Logistics Management

Security issues arise in the case of IoT applications mainly because they are often deployed on uncontrolled environments and may need to interface with hostile environments. One of the main reasons for a security breach is interconnectivity among various heterogeneous devices in case of an IoT application. The security threats can be both from the hardware devices involved as well as the software [7]. The various security challenges of IoT and big data can be visualized using Fig. 4.

The Big Data challenge includes data reliability, where we need to be sure of the data that is going to be analyzed by the sensors so that they function correctly. An IoT security device, such as an alarm system installed in a logistics warehouse can be easily compromised by eavesdropping on the radio frequency signals used for operating the alarm system. The latest research by HP Company has shown that there are more than 60% security failures at the user interfaces, 80% violate the privacy of personal records and more than 80% of IoT devices failed to mandate passwords of sufficient strength, thereby creating various security breaches [8]. The user's data is vulnerable

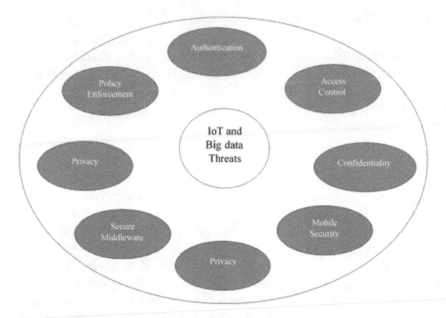

Fig. 4: IoT and Big Data Security Challenges.

to IoT devices because of their low power and security. Hence they make way for hackers to enter into the network of any logistics company. The attackers can target these vulnerable IoT devices thereby track the location of a vehicle in transit or falsify the data. IoT devices deployed to monitor and track the goods in the warehouses to enable optimal stocking can also be compromised, thereby altering the data leading to the theft of goods. Threats in the case of IoT systems can be found in the four layers which are represented in the below Fig. 5. The sensing layer consisting of physical devices and sensors, network layer consisting of transmission, internet, routing and Wi-Fi, Middleware layer consisting of API, web services, data center, and cloud and the application layer which has the various application of smart transportation [9].

The sensing layer consists of many physical sensors that may be installed at various locations in a logistics business such as warehouses, vehicles, etc; Some of the major security threats at the various layers are summarized in Tab. 2 below [9].

The IoT framework comprises of several low power nodes, such as temperature sensors, smoke detectors that may be used in logistics vehicles or warehouses. These are vulnerable to attacks as they are easily replaceable with a malicious node. Such nodes may provide access to the logistics of private data. The attackers may also install malicious code into these nodes and make them perform some undesired functions. The attacker can alternatively not hamper the code, but inject falsified data input such as stock details in warehouses, thereby leading to a denial of service attacks. There is also scope for eavesdropping attacks during data authentication and data transmission between the nodes. In sleep deprivation attacks, the battery may be drained out

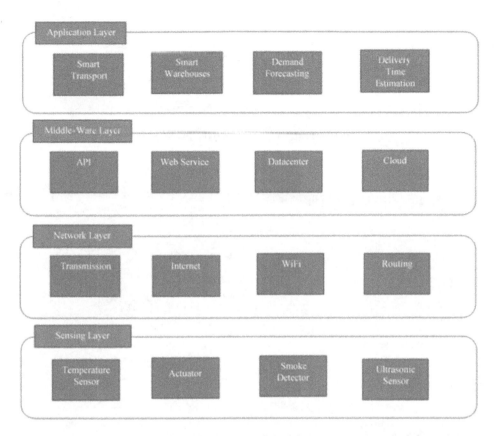

Fig. 5: Layers in IoT system for Big data logistics application.

intentionally by the attack by injecting an infinite loop that runs on these nodes until it dies. This, in turn, leads to denial of service attacks when some user tries to access the logistics system. During the booting process of these nodes i.e. when they are restarting, they are vulnerable to attacks. In phishing attacks, the attacker might lure the victim using the logistics application to visit illegal webpages, thereby compromising the user's account details such as his personal information along with passwords. In case of an access attack, the attacker can gain access to the logistics network and try to steal any confidential information of the business that can be used by the rival companies. The intention here is to only steal information and not to cause damage to the network. If the hacker can get access to the communication between the client and server, then he can become the man in the middle and control the communication without knowledge of both the parties. In the case of SQL injection attacks, the malicious queries are embedded thereby enabling the attackers to gain access to sensitive personal data of the clients stored on the database [9].

[. . .]

Tab. 2: Different Types of the attack on Big data and IoT Applications.

Layer	Security Threats
Sensing Layer	– Node Capture Attacks – Malicious Code Injection Attack – False Data Injection Attack – Eavesdropping – Sleep Deprivation Attacks – Booting Vulnerabilities
Network Layer	– Phishing Site Attack – Access Attack – Denial of Service attacks – Routing Attack
Middle Ware Layer	– SQL injection attack – Men-in-the-Middle attack
Application Layer	– Data theft attacks – Access Control attacks – Service Interruption attacks – Reprogram attacks

References

[1]   Khan, M. A., & Salah, K. (2018). IoT security: Review, blockchain solutions, and open challenges. Future Generation Computer Systems, *82*, 395–411.

[2]   Radanović, I., & Likić, R. (2018). Opportunities for use of blockchain technology in medicine. Applied health economics and health policy, *16*(5),583–590.

[3]   Joyia, G. J., Liaqat, R. M., Farooq, A., & Rehman, S. (2017). Internet of Medical Things (IOMT): applications, benefits and future challenges in healthcare domain. *J Commun*, *12*(4), 240–7.

[4]   Luo, E., Bhuiyan, M. Z. A., Wang, G., Rahman, M. A., Wu, J., & Atiquzzaman, M. (2018). Privacyprotector: Privacy-protected patient data collection in IoT-based healthcare systems. IEEE Communications Magazine, *56*(2),163–168.

[5]   Yang, Y., Zheng, X., Guo, W., Liu, X., & Chang, V. (2019). Privacy-preserving smart IoT-based healthcare big data storage and self-adaptive access control system. Information Sciences, *479*, 567–592.

[6]   Thota, C., Sundarasekar, R., Manogaran, G., Varatharajan, R., & Priyan, M. K. (2018). Centralized fog computing security platform for IoT and cloud in healthcare system. In Fog Computing: Breakthroughs in Research and Practice (pp. 365–378). IGI global.

[7]   Mahmud, R., Koch, F. L., & Buyya, R. (2018, January). Cloud-fog interoperability in IoT-enabled healthcare solutions. In Proceedings of the 19th international conference on distributed computing and networking (pp. 1–10).

[8]   Subramaniyaswamy, V., Manogaran, G., Logesh, R., Vijayakumar, V., Chilamkurti, N., Malathi, D., & Senthilselvan, N. (2019). An ontology-driven personalized food recommendation in IoT-based healthcare system. The Journal of Supercomputing, *75*(6), 3184–3216.

[9] Wu, T., Wu, F., Redoute, J. M., & Yuce, M. R. (2017). An autonomous wireless body area network implementation towards IoT connected healthcare applications. Ieee Access, *5*, 11413–11422.

[10] McGhin, T., Choo, K. K. R., Liu, C. Z., & He, D. (2019). Blockchain in healthcare applications: Research challenges and opportunities. Journal of Network and Computer Applications.

[11] Bhalaji, N., Abilashkumar, P. C., & Aboorva, S. (2019, January). A Blockchain Based Approach for Privacy Preservation in Healthcare IoT. In International Conference on Intelligent Computing and Communication Technologies (pp. 465–473). Springer, Singapore.

[12] Simić, M., Sladić, G., & Milosavljević, B. (2017, June). A case study IoT and blockchain powered healthcare. In Proc. ICET.

[13] Jamil, F., Ahmad, S., Iqbal, N., & Kim, D. H. (2020). Towards a Remote Monitoring of Patient Vital Signs Based on IoT-Based Blockchain Integrity Management Platforms in Smart Hospitals. Sensors, *20*(8), 2195.

[14] Griggs, K. N., Ossipova, O., Kohlios, C. P., Baccarini, A. N., Howson, E. A., & Hayajneh, T. (2018). Healthcare blockchain system using smart contracts for secure automated remote patient monitoring. Journal of medical systems, *42*(7), 130.

[15] Kruse, C. S., Goswamy, R., Raval, Y. J., & Marawi, S. (2016). Challenges and opportunities of big data in health care: a systematic review. JMIR medical informatics, *4*(4), e38.

[16] Bora, D. J. (2019). Big Data Analytics in Healthcare: A Critical Analysis. In Big Data Analytics for Intelligent Healthcare Management (pp. 43–57). Academic Press.

[17] Chi, M., Plaza, A., Benediktsson, J. A., Sun, Z., Shen, J., & Zhu, Y. (2016). Big data for remote sensing: Challenges and opportunities. Proceedings of the IEEE, *104*(11),2207–2219.

[18] Mehta, N., & Pandit, A. (2018). Concurrence of big data analytics and healthcare: A systematic review. International journal of medical informatics, *114*, 57–65.

[19] Nishmitha, D. S., Anvitha, A. K., & Juna Joy, S. H. (2018). Cyber physical system based health monitoring system. SYSTEM, *5*(05).

[20] Chen, M., Hao, Y., Hwang, K., Wang, L., & Wang, L. (2017). Disease prediction by machine learning over big data from healthcare communities. Ieee Access, *5*, 8869–8879.

[21] Bhatt, Y., & Bhatt, C. (2017). Internet of things in healthcare. In Internet of things and big data technologies for next generation HealthCare (pp. 13–33). Springer, Cham.

[22] Manogaran, G., Varatharajan, R., Lopez, D., Kumar, P.M., Sundarasekar, R. and Thota, C., 2018. A new architecture of Internet of Things and big data ecosystem for secured smart healthcare monitoring and alerting system. Future Generation Computer Systems, *82*, pp.375–387.

[23] Bhuiyan, M. Z. A., Zaman, A., Wang, T., Wang, G., Tao, H., & Hassan, M. M. (2018, May). Blockchain and big data to transform the healthcare. In *Proceedings of the International Conference on Data Processing and Applications* (pp. 62–68).

[24] Liang, X., Zhao, J., Shetty, S., Liu, J., & Li, D. (2017, October). Integrating blockchain for data sharing and collaboration in mobile healthcare applications. In 2017 IEEE 28th Annual International Symposium on Personal, Indoor, and Mobile Radio Communications (PIMRC) (pp. 1–5). IEEE.

[25] Hölbl, M., Kompara, M., Kamišalić, A., & Nemec Zlatolas, L. (2018). A systematic review of the use of blockchain in healthcare. Symmetry, *10*(10), 470.

[26] Karafiloski, E., & Mishev, A. (2017, July). Blockchain solutions for big data challenges: A literature review. In IEEE EUROCON 2017-17th International Conference on Smart Technologies (pp. 763–768). IEEE.

[27] Ekblaw, A., Azaria, A., Halamka, J. D., & Lippman, A. (2016, August). A Case Study for Blockchain in Healthcare:"MedRec" prototype for electronic health records and medical research data. In Proceedings of IEEE open & big data conference (Vol. 13, p. 13).

[28] Bhuiyan, M. Z. A., Zaman, A., Wang, T., Wang, G., Tao, H., & Hassan, M. M. (2018, May). Blockchain and big data to transform the healthcare. In Proceedings of the International Conference on Data Processing and Applications (pp. 62–68).

[29] Al Omar, A., Rahman, M. S., Basu, A., & Kiyomoto, S. (2017, December). Medibchain: A blockchain based privacy preserving platform for healthcare data. In International conference on security, privacy and anonymity in computation, communication and storage (pp. 534–543). Springer, Cham.

[30] Kshetri, N. (2017). Can blockchain strengthen the internet of things?. IT professional, *19*(4), 68–72.

[31] Dorri, A., Kanhere, S. S., Jurdak, R., & Gauravaram, P. (2017, March). Blockchain for IoT security and privacy: The case study of a smart home. In 2017 IEEE international conference on pervasive computing and communications workshops (PerCom workshops) (pp. 618–623). IEEE.

[32] Hammi, M. T., Hammi, B., Bellot, P., & Serhrouchni, A. (2018). Bubbles of Trust: A decentralized blockchain-based authentication system for IoT. Computers & Security, *78*, 126–142.

[33] Liu, P. T. S. (2016, November). Medical record system using blockchain, big data and tokenization. In International conference on information and communications security (pp. 254–261). Springer, Cham.

Rithvik Vasishta S, Tanmay Shishodia, Utkarsha Verma,
D Prasanna Sai Rohit, Dr. J Geetha

BECON – A Blockchain and Edge Combined Network for Industry 4.0

Abstract: With the quantity of internet-connected devices increasing and as IoT (Internet of Things) is gaining popularity, the degree of data and information collected by IoT sensors is incredibly high and needs large amounts of resources for processing and analysing the data. Edge processing permits the sensor data to be processed nearer to its source. But, resource constraints of edge computing give rise to new obstacles for regular data storage, transmission, and protection. This paper performs a survey on various implementations and researches made by professionals and institutions to analyse this innovative alternative. We also propose the utilization of a blockchain-based localized application to modify IoT edge processing. With this setup, we generate the possibility for a resource owner to hitch the system and to lend the computer resources as per requirement. The paper aims to depict the advantages of an integrated system of blockchain and edge-based offers over a network without blockchain. We aim to improve the security factor of transactions carried out on the edge network, reduce the utilization of computational power and increase scalability. The survey performed by the authors is focused on learning the advantages and drawbacks of this choice. Implementation of this framework is demonstrated using IOTA- a non-profit open-source technology organization and Tangle. Tangle is a distributed, open and scalable ledger, it supports frictionless data and transfer of value.

Keywords: internet of things, blockchain, edge computing, iota, tangle

5.1 Introduction

Internet of Things is a connected network of interrelated computing devices, mechanical and digital sensors that generate huge data for processing and analysing. With the availability of low-cost sensors around us, gathering data about our surroundings has become rather easy. Along with all the advantages of the Internet of Things, it comes with a major fallback of large amounts of data, which is ideally not easily processable.

The main concern in the IoT is the lack of security. Let us assume an example where, due to lack of security, the banking system falls into the wrong hands. It

Rithvik Vasishta S, Tanmay Shishodia, Utkarsha Verma, D Prasanna Sai Rohit,
Dr. J Geetha, Dept. of Computer Science and Engineering, Ramaiah Institute of Technology, Bangalore, Karnataka, India

https://doi.org/10.1515/9783110725490-005

would lead to a huge loss of data and money. In 2016, a major cyber-attack, in India, which compromised 3.2million debit cards crippled the cashless transaction system for a brief amount of time. In 2019, the hospital industry was one of the seriously hit victims of cyberattacks in India. FireEye, a US cyber-security company stated that Chinese-based workers hacked into the databases of Indian hospitals and leaked around 6.8million records of patient-doctor information.

To solve this problem of security we can use blockchain or any other data structure which is either based on blockchain or similar to blockchain. Blockchain is a peer-to-peer distributed data structure representing a distributed ledger. Blockchain includes a fixed of covered statistics blocks chained sequentially to one another. Together they shape an immutable ledger, allotted over the taking part nodes. Though blockchain seems to be a feasible and appropriate solution, it too has its drawbacks. It is difficult to scale, has a low transaction rate which results in high transaction fees. Protocols involved in the blockchain communication process create unnecessary overheads and increase the load. The low transaction rates of blockchain and blockchain-based software are the major setbacks currently. While blockchain can handle 6–8 transactions per second, Ethereum blockchain-based software can handle nearly 100 transactions per second, in comparison with VISA-which can handle 2000 transactions per second.

FogBus is a Blockchain-based framework for Edge and Fog computing. It aids developers and other stakeholders to run numerous applications at a time and also assists service providers to control their materials and resources effectively [1]. Blockchain has been used here to apply a layer of authentication and to ensure data integrity. But These security features take up a lot of computational power and this can impact the network usage and service delivery latency. To address the vulnerability issues of IoT, EdgeChain was proposed which integrates blockchain and smart contracts capabilities. It has its own internal currency that connects the IoT devices with the edge resource pool [4]. But it considers only some security features and does not offer any purview to app designers and stakeholders to regulate the framework according to individual requirements.

To overcome these issues of blockchain, Tangle was introduced. Tangle is a light-weight device of information of transactions that aren't collected in blocks or prepared linearly, like withinside the blockchain. Instead, it is greater like a tree, or a graph. This is advanced to blockchain due to the fact tangle is greater fluid and scalable, there is no unnecessary overhead in Tangle, lack of block rewards, and miners.

The paper is arranged in sections visiting the previous works done in the field of edge and blockchain, the challenges faced by the authors and the model envisioned. The proposed implementation of the model is explained along with the results observed. Finally, we look at other use cases where the proposed integrated network can be employed.

5.1.1 Related Work

Blockchain technology is an integration of encoding, P2P transmission, collective agreement, distributed storage, and alternative technologies. The distributed and large data traffic becomes gridlocks that retort to the needed QoS (quality of service) [14]. Knowledge accession and exchange are the basis of all the system operations. Reducing the delay of knowledge exchange to attain greater effectiveness is a challenge. Amongst the key issues of blockchain-based IoT is analysing the technology. When associated with blockchains, an industrial control system can benefit from a cost-effective and redistributed resource management that is robust against security threats. By assigning computing and storage necessities among all devices within the network, blockchain manifests a peer-to-peer network, which lowers the installation and maintenance costs of centralized cloud, knowledge centre, and network devices. This model for communication addresses the difficulty of single-point failure. Blockchain solves the drawback of a lack of privacy in IoT by using an encoding algorithm. It conjointly resolves the liableness within the IoT through the employment of ledgers that cannot be tampered with [13].

Blockchain involves mining to add new blocks to the chain. Proof of Work (POW) is a time and resource ingesting cryptographic puzzle that 'Miners' attempt to solve for mining blocks. This newly created block is added to the perpetual ledger after whichever miner solves the puzzle first. Few constraints are – delay in transaction affirmation, High computational power for determining POW, and truncated extensibility (as a copy of the block is transmitted to everyone). Furthermore, the most glaring issue with blockchain is the ineptitude of frail devices to take part in it. One of the proposed frameworks [15] involves 'Pythias', which utilizes the assistance provided by blockchains. It functions as a mediator between the blockchain and the devices and hence resolves the inability of feeble devices taking part in the blockchain. Another proposal suggests the utilization of Ethereum or Hyperledger with smart contracts on top of it. In [2] we see the proposal of a system with a disseminated edge computing layer of Fog nodes deployed among the disbursed heterogeneous IoT nodes and a centralized IoT cloud that utilizes the various benefits that fog computing provides. The network equips blockchain technology on a distributed Software Defined Networking (SDN) controller scheme. This network comprises distributed OpenFlow switches (OF) that are deployed with bounded proficiencies. Another model proposed for enhanced efficiency of edge networks was by [16], they propose the Pythia model which consists of a set of APIs to IoT devices, and these interfaces offer abilities around device identity, proprietorship, authorization for sharing data, etc.

We propose a novel implementation for the security of IoT use cases, offering full integrity, including networks created with nodes going up to thousands in number. We tend to use the underlying technology of the Tangle – a framework designed particularly for IoT and is expandable, dispersed, flexible, and free of cost. It is used in association with IOTA. To summarize, it solves each economic and quantifiability issue because it needs the user to perform a sort of labour proof that approves two dealings in a transaction. All dealings are squared with validation of the aforementioned transaction. This nullifies the need for committed miners and decentralizes the system fully. In IOTA, network transaction speed increases with an increase in the number of users.

Further we discuss a few publications made in the past on the topic under discussion. The authors performed an extensive research on various researches made by fellow explorers and attempt to interpret a few below.

The author(s) [1] introduces their research work by saying that, lately a lot emphasis is given on integrating Edge, Fog and Cloud infrastructures to aid the execution of diverse latency-touchy and computing-in depth Internet of Things (IoT) applications. To cope with the restrictions like platform independence, security, useful resource management, and multi-software execution, the authors proposed a framework called FogBus that facilitates end-to-end IoT-Fog (Edge)-Cloud integration.

The FogBus framework integrates numerous hardware units thru software program additives that provide dependent verbal exchange and platform-impartial execution of applications. It consists of Hardware units (IoT Devices, FOG Gateway Nodes, FOG Computational Nodes, Cloud Datacentres), Software components (three types of software services viz., Broker, Repository and Computational services) and network structure (to facilitate the numerous data and information shared by the software components).

The Design and Implementation of FogBus consist of Division of each system service into Service Interface and Management activity, Blockchain for maintaining the integrity of data and data prevention from tampering. FogBus gives flexibility to vendors for the usage of one of a kind custom designed or third-birthday birthday celebration Cloud Plugin offerings to combine Cloud and Fog infrastructure for computing purposes. FogBus additionally gives builders a few tips to construct their front-cease and back-cease utility packages aligned with the capabilities of the FogBus framework.

From the statistical proof given by the authors at the results, we can conclude that their framework (FogBus) is the lightest of all the other frameworks, can harness both edge and remote resources simultaneously and have the system initiation time of all frameworks.

This particular paper [2] says that Designing IoT networks face many demanding situations that consist of protection, big visitors, excessive availability, excessive reliability, and strength constraints. Fog computing is a shape of edge computing that has been evolved to offer the computing capabilities (e.g. garage and processing) at the brink of the get entry to network. A visitor's version is proposed to version and examine the visitors amongst one-of-a-kind elements of the network. The proposed paintings achieve diverse advantages to the IoT network, along with latency reduction, protection development and excessive performance of useful resource utilization.

IoT technology is always defined by the three-layer reference model. The IoT structure can be considered as a Perception layer, Network layer, and Application layer.

Edge computing is a brand-new paradigm that ambitions to offer cloud offerings and computing capabilities (e.g. garage and processing) at the brink of the get entry to network; one or hops far-far from the end-user. This introduces a manner of transferring from the centralized massive statistics centres to the allotted cloud devices with constrained capabilities. Fog computing is a shape of facet computing this is appropriate for IoT networks [20]. It introduces a brand-new computing paradigm that acts as an extension to the cloud computing paradigm capable of offer processing, computing, and garage capabilities. It additionally introduces different cloud offerings to the conversation nodes withinside the region of the allotted Fog nodes.

SDN is a brand-new paradigm that bodily separates the forwarding plane and the manage plane to offer a dynamic community structure. The Data plane represents the community component this is chargeable for forwarding traffic, even as the manage plane is the component that comes to a decision the traffic. SDN networks commonly include a centralized or disbursed controller scheme and disbursed forwarding gadgets or switches.

The authors declare that using dispensed Fog computing for IoT networks achieves numerous advantages because it brings the cloud computing competencies close to IoT nodes. This idea has brought a framework of the IoT device that deploys dispensed Fog computing with the SDN and blockchain paradigms.

Usually, [3] all the decision making and analytics take place within the Cloud thanks to their easy service-oriented access to seemingly infinite resources. Though the network bandwidth to send large data can be punitive and the roundtrip latency high. Hence the use of Edge and Fog resources as computing platforms. However, the lack of a platform ecosystem for use of these resources acts as a key hurdle. In this paper, the author proposes ECHO, an architecture, and platform implementation to address these needs. The platform must have support for Big Data applications like TensorFlow, ApacheSpark, etc. Since IoT devices are distributed, network connectivity between them is critical. Just like Cloud services, platform services are added on the edge which can manage resources and the application life cycle. Unlike Clouds, edge resources can vary with time. So, registry service is required to track their applications.

The author emphasizes the need for a middleware platform to handle data flow between Edge, Fog, and Cloud resources.

Most of the data from IoT devices are sent to the remote Cloud to be processed. But the triumphing centralized cloud computing version is extraordinarily hard to scale with the projected big range of gadgets way to the big quantity of generated statistics and consequently the especially lengthy distance among IoT gadgets and clouds. The IoT gadgets are especially inclined and might be especially effortlessly managed with the aid of using malicious hackers. To address these issues, the author [4] proposes a framework called EdgeChain. EdgeChain integrates authorised blockchain and smart contracts capabilities. It has its internal currency linking the edge cloud aid pool with IoT gadgets. Permissioned blockchain is used for greater managed and controlled surroundings and better throughput and to help allotted IoT applications. The edge servers monitor, create and append new blocks for brand spanking new transactions. It has a blockchain server that executes the clever contracts, accumulating transactions amongst gadgets. All the sports are logged at the blockchain server. The creator pursuits to combine blockchain in area computing without overloading the gadgets with full-scale security-associated burdens. He performed diverse experiments and concludes that the price to combine blockchain is inside an inexpensive variety even as gaining its benefits.

IoT devices must focus their energy in the execution of core functionalities rather than focusing on security and privacy issues – enter blockchain. The author [5] proposes a framework that relies on hierarchical structure and distributed trust for efficient blockchain security while being suitable and specific to IoT applications. He eliminates the concept of Proof of Work and the need for coins. Blockchain-based architecture induces computational and packet-based overheads. To evaluate them the author simulated a smart home scenario. He concludes the following: The packets payload size increases however it has the increase in the packet size has a relatively small effect. Energy consumption also increases however the relative increases are not significant. In conclusion, the security and privacy benefits gained outweigh the low overheads introduced by blockchain-based architecture significantly. The same architecture can be used in other domains of IoT applications as successfully as the author did in the case of a smart home.

In this paper [6], Edge Computing is introduced and its contribution To IoT networks that run Blockchain, because the software of this brings a few troubles alongside benefits. The principal problem is the constrained garage capability of IoT gadgets which might be placed in a clever, round metropolis, and the quantity of information that they should keep. To this end, Edge Computing is employed. Devices of this community can be capable of talk with Edge Computing nodes and among every other, and shop new blocks of information and transactions now no longer locally, but, on Edge Computing. In this way, it isn't essential to eat their CPU and they could use those assets alternatively. Thus, on this way, the take a look at has proven that the overall performance and performance of IoT gadgets can be increased, which may be very crucial in a real-global scenario. Furthermore, via way of means of storing information on Edge Computing nodes, the functions of records circularity in a clever metropolis idea may be fulfilled, because the extraction of information may be carried out to mitigate the assets intake and the waste manufacturing to enhance the offerings of this idea.

The structure proposed on this paper [7] lays the foundation for similarly studies on this area, imparting a lightweight, sle and personal framework that keeps maximum advantages of blockchain technology. The proposed blockchain-primarily based totally IoT structure on this version handles maximum safety and privateness threats like DOS assault, change assault, losing assault and mining assault at the same time as thinking about the resource constraints of many IoT devices. The qualitative overhead evaluation of the structure completed on this take a look at has proven that it has a regular overall performance overhead at best, and at a worst maximum of its transactions scale with the quantity of clusters withinside the community, instead of the quantity of nodes. This structure has been provided withinside the context of a clever home, but it is able to be relevant to maximum multi tiered IoT community topologies. The intrinsic broadcast medium, decentralization, and resource-constraints of IoT are key demanding situations closer to answering questions like vulnerability to a DOS assaults and the 51% assault for organising allotted trust.

The proposal offered here [8] examines key protection troubles in IoT structures with a unique emphasis on area gadgets. A business IoT area-tool the use of MQTT (Message Queue Telemetry Transport) (+TLS) that's a low-energy, low-reminiscence messaging protocol. MQTT has been extensively followed in mobile-primarily based totally messaging applications. The low-weight and low-energy traits of MQTT make it appropriate for use in limited IoT area nodes. Another area tool the use of CoAP (Constrained Application Protocol) (+DTLS) protocols have been used to research the effect of those protection concerns. CoAP is a request-reaction messaging protocol evolved for limited IoT gadgets. It implements a Representational State Transfer (REST) structure that offers it an area over different protocols. This protocol allows limited gadgets to apply net services, combining the advantages of HTTP and MQTT. By default, CoAP and MQTT protocols do now no longer use any protection layer. However, those protocols do provide the choice of more protection layers primarily based totally on TLS (Transport Layer Security). MQTT gives 3 unique protection layers, the primary of that's TLS. CoAP gives a lighter model of TLS known as Datagram Transport Layer Security (DTLS). Results of this take a look at confirmed that Neither of the alternatives become subservient to assaults like Ping of demise or malware, Introducing TLS or DTLS led to mediating assaults like sleep deprivation, packet sniffing, and node replication only, all alternatives have been nevertheless prone to sync assaults, records injection, passive reconnaissance, and malicious nodes and ultimately Code injection become now no longer viable in this area node.

In this paper [9], a machine structure that may save you statistics forgery through changing current centralized database techniques to disbursed kind primarily based totally on blockchain has been proposed. By dividing the proposed machine shape into the cloud, fog, and edge, this examine has proposed a manner to organically function the IIoT ecosystem. This paper additionally investigated if the overall performance of the block-chain community settlement set of rules may be assured through the usage of public cloud resources. This examine resolves statistics validation and protection primarily based totally at the Hyperledger blockchain platform withinside the IIoT environment (legal blockchain), and additionally resolves overall performance load problems through now no longer connecting IIoT gadgets immediately to block-chained networks, Orderer is ported to the cloud to make certain stability, protection, and scalability. It proposes that clever agreement and transaction verification continue to technique in fog node to gain community latency and throughput overall performance. Challenges consist of studies into community configuration for price financial savings withinside the cloud and optimization of cloud and fog node instancing overall performance to maximise the overall performance of the block-chain community Hyperledger and to investigate the overall performance and consistency of the version through making use of it to the real enterprise environment.

This article [10] explores the suitability of edge computing for rising IoT packages. Specifically, the paper evaluates the overall performance of part computing for cell gaming as a consultant state of affairs of recent packages incorporating bodily sensory inputs further to the ones explicitly generated through the user. It is a regarded reality that the quantity of information generated via way of means of sensors, actuators and different gadgets withinside the Internet of Things (IoT) has drastically multiplied withinside the previous few years. IoT information are presently being processed withinside the cloud, generally thru computing assets placed in remote information centres. This reasons network bandwidth and conversation latency to come to be critical bottlenecks. Also, the growing quantity of statistics exchanged provides extensive pressure at the network hyperlinks to the cloud. This article advocates aspect computing for rising IoT packages that leverage sensor streams increase interactive packages. The Paper demonstrates an experimental assessment of aspect computing and its permitting technology in a particular use case represented with the aid of using cell gaming.

The paper explores the idea of bringing content closer to the end-users. For this the CDNs (Content delivery or distributed networks) are proposed, those install assets that mirror content material from a supply place onto servers near the end-users. Information-centric networking (ICN) is a comparable method for reinforcing the Internet infrastructure to explicitly assist content material-primarily based totally routing and forwarding. On the alternative hand, facet computing servers additionally offer computational abilities and might host interactive packages that assist person mobility. Furthermore, an facet computing platform can relieve privateness issues because the statistics generated from IoT gadgets are saved and processed inside nodes withinside the facet network.

The paper classifies the proposed edge architectures into three categories- resource-rich servers (deployed close to the end-devices), heterogeneous nodes at the edge (including the end-devices themselves), Edge Cloud federation (centralized data centres). The article additionally discusses numerous unique capabilities of the IoT that make it suitable for deployments primarily based totally on facet computing platforms, namely- low latency comms, require higher bandwidth standards, geographical dispersal, and tool mobility.

Finally, the use case of mobile gaming selects Neverball evaluation. Neverball is consultant of a bigger magnificence of programs that depend upon rendering complicated 3-D environments, along with digital and augmented reality. The effects of the test encompass the effect of server deployment on community delay, the overhead of various virtualization technology and the effect of extra computational assets provided via way of means of the cloud at the processing delay.

The paper settles with evidence of facet computing being essential to allow quick interactive games. Though nearby information centres permit to significantly lessen community latency, the simplest web hosting sources at the brink permits a first-class nice of enjoy for gaming. The paper establishes that deploying even confined computing sources at the brink facilitates enhance the nice of enjoy.

A key challenge in the deployment of blockchain technology is the hosting location. This paper [11] evaluates the use of the fog and the cloud as possible platforms for edge devices. Blockchain is a decentralized ledger that contains connected blocks of transactions. Due to its distributed and decentralized organization, blockchain is being used within IoT. The capacity to create /store /switch virtual property in a distributed, decentralized and tamper-evidence manner is of superb sensible price for IoT systems. A key task withinside the deployment of Blockchain as a Service (BassS) for IoT is the web website hosting environment. While the fog has constrained sources, it famous low latency. Edge gadgets are regularly too restricted concerning computational sources and to be had bandwidth main to cloud or fog as cap potential hosts. To compare the usage of fog, a easy test turned into conducted.

In conclusion to all the studies analysed above, one can say that there exists a variety of applications that deal with the security of edge devices over a network. Each study has its advantages and challenges, and all the studies focus on improving their product as future prospects. All the papers discussed, have used blockchain as a measure to improve the security of their application against threats like DOS attacks, modification and mining attacks, etc., or for improving the efficiency of their application through decentralization feature of blockchain as this offers every single user an opportunity to become one of the payment processors in the network, other key features of blockchain have also been implemented in most of the studies. We aim to refer to these studies to analyse how blockchain and its key features function in fog computing and use these methodologies in our product to improve its security, and to improve upon the functionalities and challenges that other studies could not implement in their products.

5.2 Motivation and Challenges

5.2.1 Motivation

One of the pressing issues faced in the IoT domain is the security of the IoT devices across the network. We try to address these issues in our proposed framework which comply with the following goals:

5.2.2 Decentralization

A majority of the IoT deployments are on the cloud, i.e., a central storage facility and data processing architecture is followed. This means that the participants of the deployment have to send data to the cloud to store/process and wait for the cloud processor to complete its task then to send it back to the participant. Naturally, this process becomes cumbersome and hectic for the cloud processors/storage unit to handle when the number of participants or the data to be processed increases. The emergence of a new networking philosophy as edge computing reduces latency and usage of bandwidth because it brings the computing and storage as close to the source as possible. Also, the absence of single-point processing/storage reduces the risk of failure across the system and increases scalability and robustness.

5.2.3 Security

As mentioned earlier, one of the most pressing issues in IoT is across-system security. Generally, device identification is not performed, Data encryption is not done. In an ideal and effective framework, the device identity has to be verified and cross-checked with the registered devices and all the data which flows through the network has to be encrypted and maintained for easy and quick access.

5.2.4 Challenges

Initially, we tested our framework on an open-source, blockchain-based [16] decentralized software platform and found out the following challenges:
– Miners in blockchain attempt to solve a resource-intensive problem called 'Proof-Of-Concept' after which, it adds a new block to the present chain. Since this process of 'mining' is resource-intensive, many IoT devices cannot afford the required resources.
– Mining in the blockchain is time-consuming and since mining, the basic building block in the blockchain-based network high latency is an obvious by-product, whereas low latency is desired in most of the IoT deployments.
– Since Blockchain usually deals with cryptocurrency, the unnecessary protocols in blockchain-based software create overhead, which is unnecessary for IoT deployments.

5.3 Model

A high-level view of IOTA:

IOTA is a distributed ledger that overcomes the limitations of Blockchain and serves as the foundation for IoT and M2M applications. Transactions in IOTA are efficient, much more secure though not as much as Blockchain, very lightweight, and require no miners in real-time. It's a cryptocurrency that's not built on blockchain like any other cryptocurrencies out there and has its own data structure called tangle.

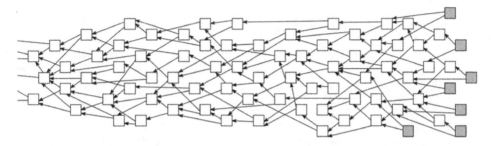

Fig. 1: Tangle Data Structure.

The tangle, unlike forming a chain, makes use of a Directed Acyclic Graph (DAG): a directed graph with no formation of directed cycles (Fig. 1). Hence it is also known as the block-less blockchain. It retains the distributed ledger's blockchain features, in which a mesh of independent accounts performs transactions among themselves, reaching consensus about who owns what without depending on a centralized authority. The IOTA tangle has the ability to create microtransactions known as Smart Contracts using scripting languages. Tangle, cryptocurrency, and Smart Contracts together enable an environment where accounts or devices can share in the system.

The following features of tangle make it well suited for IoT applications.

1. **Decentralization:** There is no central control that removes a single point of failure and makes the system more robust.
2. **Scalability:** The number of transactions in the system increases the rate of realization of transactions. More transactions can be confirmed as the transactions occurring increase since each transaction requires the sender to verify two other transactions on the Tangle network. This means that IOTA scales proportionally to the number of transactions. For confirmation of these two transactions, the device performs low difficulty "proof of work", which is essentially just a series of mathematical problems.
3. **Fast transactions:** For the same above reason, more the number of transactions less will be the time taken for each transaction. Since transactions will get validation from other nodes.

4. **No concept of fees:** Since there is no group of distinct miners that validate the transactions there are no fees. Though validation requires proof of work so it's not strictly free.
5. **A fixed money supply:** Every IOTA token that is created exists in the genesis block itself. The total money supply or IOTA tokens are 2,779,530,283, an amount which will never increase or decrease.

5.4 Implementation

An IOTA transaction involves the following main steps:
1. **Making the transaction bundle and finalizing:** The unit of transaction is a bundle that includes 3 types- input, output, and meta transactions. We create balanced bundles that are finalized using kerl hash functions to get the bundle hash.
2. **Signing/encrypting transaction:** The input transaction is signed with the corresponding address's private key.
3. **Getting two tips – trunk and branch:** Two pending transactions in the tangle are randomly selected with the help of the Markov chain Monte Carlo algorithm and authenticated.
4. **Proof of work:** In one bundle, all will have the same tips. All transactions in the bundle from the last index to 0 are filled with the trunk, branch hash, timestamp, and then do PoW to find nonce. Transaction hash is generated and validated.

5.5 Proposed Implementation

We propose a high-level implementation of communication between two edge devices in an IOTA network. An IOTA enabled car requests for a charger to start charging and only when the access is authenticated the charger starts charging. Let us assume that the car and charger are not connected to each other at the start.

In this implementation, we integrate the following functionalities:
1. Start charging (Both smart car and smart charger)
2. Stop charging (Both smart car and smart charger)

The Steps to configure the Tangle network to coordinate with the mobility platform are as follows:
a) Connect to the IOTA community network using the API provided by IOTA (compose API).
b) Encrypting a unique seed using sha256, which makes it hard to hack by brute-force. The seed is 81 characters long and is basically used to calculate the address of the which are used by the different edge devices on the network to

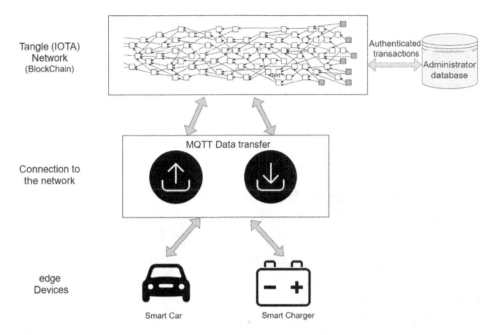

Tangle (IOTA)
Network
(BlockChain)

Authenticated
transactions

Administrator
database

MQTT Data transfer

Connection to
the network

edge
Devices

Smart Car

Smart Charger

Fig. 2: A brief architecture diagram for our implementation.

inter-communicate. Generating a seed is necessary because it always generates
the same address which is better than storing a private key-address on the database making it vulnerable to attackers (fig. 2).

c) Generating the address from the aforementioned seed using the IOTA API (get-
NewAddress). This address is used to identify uniquely our edge devices on the
tangle/IOTA network. All the transactions are carried out via the addresses allotted to the components in the network.

d) Recharging the components with a sample amount of IOTA token will allow us
to make the transactions happen as a normal payment system.

The Steps to connect the components (Smart Car and Charger) from the high mobility community platform to the Tangle network:

a) Initialize IOTA enabled electric car and charger using HMKit with their seeds
and access tokens. The car should have enough IOTA tokens as described in the
above steps to do payment.

b) The two components (car and charger) are encapsulated using classes which
include the necessary function like initializing the HMCertificate, startCharging,
sendPayment, authenticate, authorize, etcetera.

c) The car issues a charging request at the address of the charger with the required
payment value (sendTransaction).

d) The transaction is sent as a transaction bundle. (sendIOTA)

e) The charger verifies whether the payment has been confirmed or not. A transaction hash is generated. Only after confirmation of payment, the charging of the vehicle is initiated.

f) We can see the starting of charging on the high mobility community platform and also from the deduction of iota tokens at the car account and addition of iota tokens at the charger account.

5.6 Different Scenarios in the Implementation

In our implementation, we have addressed the following cases along with how the system reacts to the respective scenarios.

5.6.1 Case 1: Single-Car Single-Charger

In this scenario, the car and charger are initialized with the address, which was derived from the respective seeds. Assuming the car has been topped up with enough iota tokens, it sends a charging request to the charger. The following results can be expected:

a) **The IDs of both car and charger are matching (from the admin database):** This is the ideal case where the car and charger are ready to transact. Once the transaction begins and charger authenticates the payment, the car-charger ecosystem starts charging

b) **The IDs of the car and charger do not match:** In this case, the charger will reject the transaction request and wait for a new request.

c) **Insufficient tokens at the car:** When the IDs match, there might be a possibility that the iota token in the car might be not sufficient for the transaction. Hence the charger will issue an 'insuffientAmount' error and reject the request.

5.6.2 Case 2: Number of Chargers is Greater Than That of Cars

In this scenario, the chargers are in excess. All the cars can be accommodated with a charger and the transactions can take place as usual. Once the cars and chargers are coupled the above-mentioned scenarios can be expected.

5.6.3 Case 2: Number of Cars is Greater Than That of Chargers

This is a challenging scenario where the number of cars is greater than the number of chargers. Initially, the cars get matched with the available number of chargers and initiate charging. Again, any of the aforementioned scenarios can be expected. Once a charger is free (either finish charging or reject a transaction) it can get matched with a car which queued waiting for its turn to initiate the transaction. This particular scenario takes longer than a single cycle charging because the charging process in our implementation happens serially and not pipelined.

5.7 Problems Addressed

a) **Scalability:** Although our IOTA network contains very few devices, more the number of devices, more transactions get validated. Whereas increasing the number of users in a blockchain network would add more overhead.
b) **Separation Tolerance:** In real-time applications like this scenario, devices should be able to disconnect from the Tangle network as and when required. They can later reconnect.
c) **Quantum Proof:**Quantum machines are the next generation of computers which can run algorithms at lightning speeds and have the potential to crack any encryption. Cryptocurrencies aren't safe to these threats even though IOTA is resistant to QCs for now as it uses hash-based signatures.

5.8 Results

We conducted a comparative experiment for our implementation with both Ethereum (a third party blockchain-based service) and Tangle (IOTA) and the following observations were made.

Parameters	Blockchain/Ethereum	Tangle(IOTA)
Number of transactions per second	7–9 transactions per second	500–800 transactions per second
Time of execution	116ms (for the single-car single-charger scenario)	9ms (for the single-car single-charger scenario)
Security concerns	As blockchain is well developed and tested, it is very secure but it is prone to be attacked by the 51% attack.	Tangle (IOTA) is a well -connected DAG, once it is fully connected it becomes very secure but as tangle is a newly developed data structure it is more prone to attackers.

(continued)

Parameters	Blockchain/Ethereum	Tangle(IOTA)
Support for edge (IoT) devices	Since Ethereum is a network-heavy, it's support for edge devices is very limited because of the time constraint	Tangle can support an enormous number of transactions between the edge devices since it is a network of connected idevices that can share data.
Fees	0.001 mBTC /kB(minimum).	As the idea of Miner is non-existent in Tangle (IOTA) there are no extra fees which must be paid.

Based on the observations, we can infer the following things:

1. Tangle is very fluid and scalable whereas blockchain is very rigid and hinders scalability.
2. Tangle takes significantly less time for the execution of our implementation than blockchain which means Tangle (IOTA) is the fastest among itself and blockchain.
3. Since both blockchain and tangle have their own security flaws, blockchain can be considered to be a more secure blockchain as it has been around for a long time and has been very well developed whereas tangle has a lot to develop.
4. It has been theoretically proven that tangle is the best choice for supporting IoT devices than blockchain.
5. As blockchain's basic building block is mining, a lot of fees has to be paid for the miners whereas tangle does not have any miners there are no extra fees to be paid.

According to these inferences, it is clear that tangle is the best choice for our implementation and have successfully implemented a system of a car and charger based on edge and tangle.

The prowess of IOTA Tangle compared to Blockchain can also be represented diagrammatically using a simple mathematical graph after conducting experiments based on both of them. The graph is a comparison of the capacity of the network and the number of transactions that the network can handle.

The graph clearly depicts how Blockchain struggles with increasing transactions whereas Tangle gets better with no restricted number of transactions that are to be made per second. From all of the above observations, it is proved that the usage of Tangle-based IOTA network provides security for the transactions, requires comparatively less computational power and is easily scalable. Even though it may take a longer time for execution, the security advantage overrules the time constraint.

Drawing from these results, it can be analysed that a resource owner can allocate resources to a computer as per the needed constraints, this would give the IoT nodes some amount of leverage to dispense a part of the processing load to the resource owner nodes if required. From the implementation proposed and the experiment conducted, it can be concluded that BECON is a viable blockchain-based localized application for IoT processing (fig. 3).

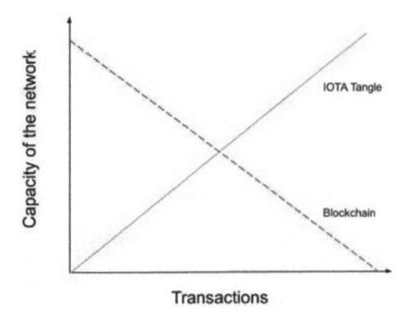

Fig. 3: A simple comparison between Tangle and Blockchain.

5.9 Use Cases

IOTA is a non-profit open-source technology organization and is the inventor of the Tangle, a distributed ledger, and protocol for the Internet of Things. IOTA solves Blockchain's problems of scalability, energy requirements, data security, and transaction fees. The Tangle is a directed acyclic graph (DAG), and doesn't involve any blocks, mining and is free of cost. Apart from the Mobility use case implemented in this paper, involving cars and chargers as the edge devices communication through the tangle blockchain to identify it right charger to charge a car, the other use cases that can employ IOTA and Tangle are – Supply Chain and Global Trade, Industrial IoT Ehealth, Smart Energy and Environmental Impact.

5.9.1 Industrial IOT

An IOTA-enabled smart factory logs commands & an encrypted hash is securely stored in the Tangle. Only the hash is stored in the Tangle, creating a tamper-proof audit trail

5.9.2 EHealth

Giving the power of controlling their own health into the people's hands. IOTA secures patient data & democratizes access. IOTA (MAM) protocol for remote patient monitoring secures data streams from remote sensors using healthcare interoperability standards. MAM also allows for granular and secure data exchange between healthcare providers, citizens, and caregivers.

5.9.3 Smart Energy

IOTA enabled decentralized Peer-to-Peer energy trading and autonomous "flexibility" markets can shape self-sufficient positive energy communities. Seamless EV Smart Charging based on Plug & Charge / M2M micro-payments. Sustainable energy traceability and REC tradeable certificate provenance Remote control of energy assets and Demand Response system for grid stability/peak shaving.

5.9.4 Environmental Impact

Losing significant species of botanical life to deforestation has cost the environment a great deal. Though carbon credits are used to track environment-friendly purchases/investments, they are hard to track and easy to duplicate or create fake ones. Now simple sensors can detect and monitor fires or trees being cut. CO_2 emissions can be precisely measured and evaluated. The data stored in the Tangle allows real-time action

5.9.5 Humanitarian Impact

The unfortunate aftermath of any kind of shift in governing power is that there are billions of people that end up trapped in refugee camps for years, simply due to missing paperwork. "Unified Identity" requires only your own body to ensure the human right to personal identity. Through a palm vein scan, retinal scan, or other biometric identification, people can be identified quickly and easily. IOTA's distributed ledger technology allows records of health and identity, work performed in camps, & educational credentials to be securely validated, providing an immutable audit trail, never to be manipulated and only seen by those granted access.

5.10 Conclusion

IoT and Edge computing have evolved and have improved data transmission and computation scenarios in the industry. Edge computing optimizes net gadgets and net packages via way of means of bringing computing towards the supply of the data. This minimizes the want for long-distance communications among customer and server, which reduces latency and bandwidth usage. Though edge has its benefits, it also brings with its drawbacks like weak authentication, insecure communication, and risk of service interruptions. These drawbacks are overcome by using a blockchain, a dispersed record maintained by a P2P network, which is a cluster of nodes, on top of edge networks to increase security in data transmissions. Information encrypted into a blockchain is immutable and protected against tampering. Blockchain protects transactions from being modified or removed, records an immutable history of events, and offers complete transaction traceability for users.

Our implementation of a mobility system with help of IOTA and Tangle is a clear explanation of how edge devices communicate over a blockchain framework (Tangle, in this case) and how all communications are protected by hashes and no tampering is allowed. The cars and chargers that form the edge devices in the system recognize their compatible counterpart by matching their respective hashes and if these do not match, a flag is raised and the transaction does not take place.

We establish the fact that blockchain improves the functionality and efficiency of edge computing networks.

References

[1]   Tuli, Shreshth & Mahmud, Md & Tuli, Shikhar & Buyya, Rajkumar. (2018). FogBus: A Blockchain-based Lightweight Framework for Edge and Fog Computing.
[2]   Muthanna, Ammar & Ateya, Abdelhamied & Khakimov, Abdukodir & Gudkova, Irina & Abuarqoub, Abdelrahman & Samouylov, Konstantin & Koucheryavy, Andrey. (2018). Secure IoT Network Structure Based on Distributed Fog Computing, with SDN/Blockchain.
[3]   Ravindra, Pushkara & Khochare, Aakash & Reddy, Siva & Sharma, Sarthak & Varshney, Prateeksha & Simmhan, Yogesh. (2017). ECHO: An Adaptive Orchestration Platform for Hybrid Dataflows across Cloud and Edge.
[4]   J. Pan, J. Wang, A. Hester, I. Alqerm, Y. Liu and Y. Zhao, "EdgeChain: An Edge-IoT Framework and Prototype Based on Blockchain and Smart Contracts," in IEEE Internet of Things Journal, vol. 6, no. 3, pp.4719–4732, June 2019, DOI: 10.1109/JIOT.2018.2878154.
[5]   A. Dorri, S. S. Kanhere, R. Jurdak and P. Gauravaram, "Blockchain for IoT security and privacy: The case study of a smart home," 2017 IEEE International Conference on Pervasive Computing and Communications Workshops (PerCom Workshops), Kona, HI, 2017, pp. 618–623, DOI: 10.1109/PERCOMW.2017.7917634.

[6]   A. Damianou, C. M. Angelopoulos and V. Katos, "An Architecture for Blockchain over Edge-enabled IoT for Smart Circular Cities," 2019 15th International Conference on Distributed Computing in Sensor Systems (DCOSS), Santorini Island, Greece, 2019, pp.465–472, DOI: 10.1109/DCOSS.2019.00092.

[7]   A. Stanciu, "Blockchain-Based Distributed Control System for Edge Computing," 2017 21st International Conference on Control Systems and Computer Science (CSCS), Bucharest, 2017, pp.667–671, DOI: 10.1109/CSCS.2017.102.

[8]   S. Shapsough, F. Aloul and I. A. Zualkernan, "Securing Low-Resource Edge Devices for IoT Systems," 2018 International Symposium in Sensing and Instrumentation in IoT Era (ISSI), Shanghai, 2018, pp.1–4, DOI: 10.1109/ISSI.2018.8538135.

[9]   Jang, Su-Hwan & Guejong, Jo & Sangmin, Bae. (2019). Fog Computing Architecture Based Blockchain for Industrial IoT. 10.1007/978-3-030-22744-9_46.

[10]   G. Premsankar, M. Di Francesco and T. Taleb, "Edge Computing for the Internet of Things: A Case Study," in IEEE Internet of Things Journal, vol. 5, no. 2, pp. 1275–1284, April 2018, DOI: 10.1109/JIOT.2018.2805263.

[11]   M. Samaniego, U. Jamsrandorj and R. Deters, "Blockchain as a Service for IoT," 2016 IEEE International Conference on Internet of Things (iThings) and IEEE Green Computing and Communications (GreenCom) and IEEE Cyber, Physical and Social Computing (CPSCom) and IEEE Smart Data (SmartData), Chengdu, 2016, pp.433–436, DOI: 10.1109/iThings-GreenCom-CPSCom-SmartData.2016.102.

[12]   O. Alphand et al., "IoTChain: A blockchain security architecture for the Internet of Things," 2018 IEEE Wireless Communications and Networking Conference (WCNC), Barcelona, 2018, pp. 1–6, DOI: 10.1109/WCNC.2018.8377385.

[13]   Xu X, Zeng Z, Yang S, Shao H. A Novel Blockchain Framework for Industrial IoT Edge Computing. Sensors (Basel). 2020 Apr 7;20(7):2061. DOI: 10.3390/s20072061. PMID: 32272555; PMCID: PMC7181142.

[14]   Fremantle, Paul & Aziz, Benjamin & Kirkham, Tom. (2017). Enhancing IoT Security and Privacy with Distributed Ledgers – A Position Paper. 10.5220/0006353903440349.

[15]   Naresh E. and Vijaya Kumar B.P. 2018. Innovative Approaches in Pair Programming to Enhance the Quality of Software Development. Int. J. Inf. Comm. Technol. Hum. Dev. 10, 2 (April 2018),42–53. DOI:https://doi.org/10.4018/IJICTHD.2018040104.

[16]   Naresh, E., Kumar, B. P. Vijaya Kumar., Niranjanamurthy, M.,; Nigam, B. (2019). Challenges and issues in test process management. Journal of Computational and Theoretical Nanoscience, 16(9),3744–3747.

Abhishek K L, M Niranjanamurthy, Yogish H K

Internet of Things Enabled Health Monitoring System for Smart Cities

Abstract: Prior to Internet of Things (IoT), interactions of patients with doctors or physicians had been restrained to visits, textual and tele communications. There was no chance doctors or physicians could monitor patients' health constantly and give advice accordingly. IoT-empowered devices have made remote screening in the healthcare sector promising; unleash the ability to keep patients protected and healthy and engaging doctors to convey the standout care. It has additionally accelerated patients meeting and delight as communications with physicians have come to be simpler and steadily proficient. Moreover, remote screening of patients' health helps to diminish time-span of hospital stay and stops re-admissions. IoT additionally has a major role in reducing healthcare cost considerably and improving remedy outcomes. IoT is absolutely remodelling the healthcare sector via reanalyzing the space of devices and individuals' communication in conveying healthcare solutions. Applications of IoT in healthcare which benefit patients, families, physicians, hospitals, and physically challenged persons have been discussed in this chapter.

Keywords: Rpi, GSM, ZigBee, RFID seniors, Google API, LCD, IoT

6.1 Introduction

Nowadays technology has been updated and grown to such an extent that any and every device can be accessed from any remote area across the world. It's not efficient to have each and every device to be installed with a super computer for computing or processing and a storage system (like Database, Database Management System, etc.) to achieve mobility of devices. So to overcome this problem each and every device is interconnected in a form of network.

The IoT nowadays is widely used in many areas which make the work, collection and transfer of data instantly to the prescribed person at instant time.

Abhishek K L, Department of MCA, Ramaiah Institute of Technology, Bangalore, India,
e-mail: abhishekmelkote@gmail.com
M Niranjanamurthy, Department of MCA, Ramaiah Institute of Technology, Bangalore, India,
e-mail: niruhsd@gmail.com
Yogish H K, Department of ISE, M S Ramaiah Institute of Technology, Bangalore, India,
e-mail: yogishhk@gmail.com

https://doi.org/10.1515/9783110725490-006

This model of connecting devices to perform a specific activity is called IoT. IoT is a new revolution in the capabilities of endpoints that are connected to the internet and is being driven by the advancement in capabilities in sensor networks, mobile devices, wireless communications, networking and cloud Technologies. The scope of IoT is not just limited to just connecting things such as devices, appliances, machines etc., to the internet. IoT allows these things to communicate and exchange data while executing meaningful applications towards a common user or machine goal.

In healthcare and also in aiding physically challenged people, IoT has played a major role. Devices like wheelchair, health monitoring etc., are installed with sensors and those are connected to other devices in IoT [3]. These technologies have nullified all types of disabilities and challenges faced by people. Where the devices for monitoring and supporting are connected, the Physician or the doctor can monitor the patient's status and provide support to them. For example, IoT devices like 3D printed prosthetics helps people who are completely paralyzed. Using Amazon Alexa and other voice commanding technology [1] it is possible to create a smart home which also helps paralyzed people by which they can do an action using just their voice.

Prior to IoT the physically challenged persons are dependent on others in order to move freely from one place to another. This drawback is eliminated as IoT enabled wheelchair empowers physically challenged persons to move freely with less dependency on others [2]. It is not at all an easy task for the individuals who have foot disabilities, to carry out usual tasks in their daily life. It is very important to understand the problems of physically disabled individuals in detail and based on the details sensors should be equipped to transfer the real-time data about the physically challenged person's health status in case of emergency to the nearest hospitals. It is also possible to integrate voice controlling or assisting systems to the existing prototype. Infrared sensor based glasses are integrated to the wheelchair [6] for easy movement. These two features increase the mobility level of the wheelchair to the greater extent which makes the system more efficient.

6.2 Proposed System for Smart Healthcare Monitoring using IoT

The distinguishing characteristic of IoT in the healthcare domain is the regular monitoring of a patient by examining various parameters and infers a good output from the past of such regular monitoring.

There are many sensors available which are used to check the BP, EEG, temperature, heart beat [2]. The data which has been recorded is sent to the patients' personal doctors, hospital and family members. So that they can monitor patients' health from time to time without visiting him frequently. Additionally, there may be obstacles in offering the information and data to the pro specialist doctors and the concerned relatives and family members.

6.2.1 Sensors Connection in Smart Healthcare System

The health monitoring system consists of a Raspberry pi (Rpi) board to which the sensors and devices to measure health parameters are connected. As shown in Fig. 1, the first column blocks contain the sensors which are used to detect the reading of different parts of the patient's body. Various Health monitoring sensors are connected to the Raspberry pi using GPIO pins to create a route to exchange the information between sensors and Raspberry pi. Sensors used in Health Monitoring system are:
– Temperature sensor: is used to detect the current body temperature.
– Heartbeat sensor: it will detect the number of heartbeats per min (bpm).
– Vibration sensor: this will sense if there is any vibration shivering in the body.
– Bp sensor: this detects the blood pressure in millimetres of mercury (mmHg).

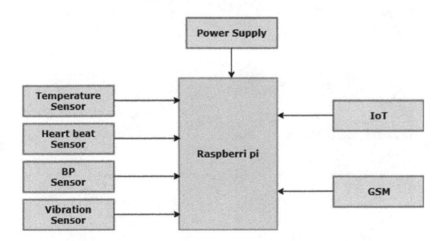

Fig. 1: Sensors connection in Health monitoring system using IoT.

6.3 Components of Smart Healthcare System

– GSM module
GPRS/GSM quad band module is used which provides GPRS connection to the system and has SIM900 communication module from SIM Com. At any point of time

the patients can log in to check and monitor their health status from anywhere. The system is smart enough to trigger a SMS/EMAIL alert. The text is sent to the family and the doctor and the nearby hospital for the ambulance so that they can take the immediate precaution for the person in case of emergency.

– Role of IoT

Now a day's IoT is playing a very important role in the day to day life. In the health-care field, IoT has made things easy to send the information about the patients' health instantly to the physicians or doctors. The below Fig. 2 describes overall components and processes involved in IoT to monitor patients' health status.

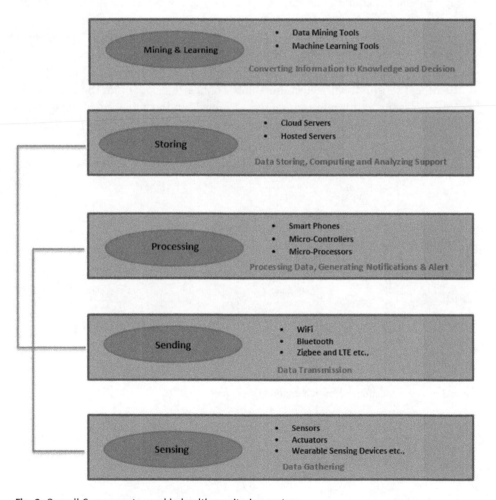

Fig. 2: Overall Components used in health monitoring system.

– Sensing

This is the first stage where all the data is gathered using sensors, actuators and wearable sensing devices which are placed in the different parts of the patient's body. Using these sensors, the readings like heartbeats, blood pressure, temperature and vibration of the body are measured.

– Sending

In this stage the recorded data of the individuals like BP, EEG, temperature and vibration of the body etc., from the sensors. These sensors are placed in the different parts of the human body and sensed data are communicated to the concerned doctor, family members and hospitals to store the records about the present condition of the person. The recorded data is sent to the server or external devices using Wi-Fi, Bluetooth or ZigBee protocols.

– Processing

In this stage the recorded data is processed based on the user requirement to get the useful information. Based on these information notifications are generated, alert messages and the location of the patient is sent to the concerned people in case of emergency.

– Storing

In this stage, data collected from the patients are stored using a cloud infrastructure which helps physicians or doctors to access the data whenever and wherever required and also to analyze the recorded data to treat the patients. The analysed data and the precaution given to the particular patient are also updated regularly in cloud servers or hosted servers.

– Mining and learning

In this stage the analysis and prediction will be done on the recorded data. Analysis on the disease can be made by the physicians or doctors using patients' data which are stored in the cloud. This analysis helps the doctors to give precautions on the particular disease and this data could be used for further analysis and diagnose the situation. Machine learning tools and data-mining tools are used to make decisions and predictions on the recorded information [4].

6.4 Proposed Smart Healthcare System for Specially Challenged People

In the healthcare domain IoT is playing a vital role and hearing aid application using IoT is already familiar.

Around 450 million people have progressive deafness problems. In the past days the hearing aids devices are simply to amplify the sound based on the hearing capacity [15]. Simply consider how the present hearing aids are no longer restricted to just amplifying sound, they have progressed and we can connect those devices to the mobile and control the amplitude of the device.

The following technologies can be used in developing smart hearing devices:
– Smartphone connected to hearing aid device.
– Voice search.
– RFID seniors.
– Biometric analytic.

These technologies have started combining and merging into one healthcare product aimed at the hard-of-hearing market. Since hearing starts to be a partial problem in humans as early as after the age of 25, it is important to bring a younger population into more likely than Baby Boomers to not only see the beneficent advances technology can bring, but capitalize on them [13].

6.4.1 Benefits of Hearing Devices

– Ability to remember the sound when the person is travelling to different locations, in restaurants, driving cars, and having group discussion.
– Hearing aids can connect to Bluetooth and devices that emit audio, like smart phones, laptops, tablets, and the like can stream the sound directly into your hearing aid without the need for headphones or removal of your hearing aid [5].
– Voice enabled personal assistants can help individuals take care of administrative tasks such as booking an appointment, a ride, song, a restaurant, a hotel, or an airplane seat [9].
– Voice search alert.

Going to the fundamental part of the IoT gadget which is work on Google speech API based aid for visually impaired individuals and deaf, speech on microphone is sent to Google API server that changes over converts into the textual form and display it on respected device which may be mobile, LCD device or some other display device and it read the text and amplifies the sound through the speakers of the device which we select [10]. This is more useful for blind and deaf people to work efficiently in the organization.

The present technique separates the required sound from complicated background sound and unwanted sound with the help of acoustic moulding module to get the desired output [4]. The present device uses the Digital Signal Processing, Pulse width Modulation and Fast Fourier Transform, which improves the quality of the device and also uses a webcam that helps to read the colour, text and converts it to the speech. The device is very helpful for both deaf an d blind people, uses the touch sense to understand the signals.

Speech from the human is generally a sound wave which travels through air varying person to person. 0.125 KHz to 12 KHz [11] is the normal frequency range for sound waves. Speech signal which is an analog one is converted to text and this signal is amplified using the client server model. Using a remote Google API server, text output for the speech is given. The sound is converted into electrical signal by microphone. And the sound is recorded which is coming from all the directions and it removes unwanted noise if present.

The filtered output from the microphone is given to the raspberry Pi which applies the corresponding libraries and speech recognition technique in order to convert the stream signal into textual format as shown in Fig. 3.

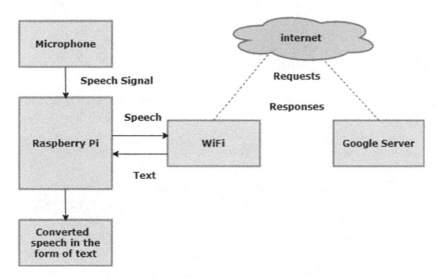

Fig. 3: Block diagram of the system for blind, dumb and deaf using IoT.

6.4.2 Hardware Description

Here the microphone, LCD and loud speaker is connected to a Raspberry pi which acts as a processing unit as shown in the Fig. 3. In order to connect to the internet, Raspberry pi also has an inbuilt Wi-Fi module. The LCD is connected to 6 GPIO pins of the Raspberry Pi to exchange the information.

The textual content which is received from the Google API server is displayed using LCD. Hence the above system can be used by the blind people to hear the speech and for persons suffering from mild deafness can see the speech converted textual data on LCD display.

6.4.3 Client Side Operations (Raspberry Pi)

Step 1: A speech is taken from microphone and unwanted noise if any is removed and sent to speech synthesizer.

Step 2: Check the speech validation using Rpi.

Step 3: The encoded audio file which is in MP3 Format is converted into NDR format

Step 4: Establish Internet connectivity Using Wi-Fi module.

Step 5: HTTP protocol is used to send the post command to Google API server

Step 6: Issue the Remote procedure protocol using Rpi's client library.

Step 7: A stub procedure is called that uses the common memory space of Rpi.

6.5 Smart Contact Lens

Smart contact lenses are currently the most awaited IoT technologies based for medical care. The smart contact lenses are similar as the ordinary contact lenses apart from their technological difference i.e. sensors and their minute processors. These contract lenses are currently used to track the blood glucose level [7] with the help of the user's tears and also to improve the vision.

These contact lenses will be used as an incredible interface between a person and an electronic device for healthcare improvisation. Smart contact lenses are basically built on a biocompatible polymer, flexible and ultrathin circuits and microprocessor chip [19] for the functioning of the contact lens. And also drug controlled delivery, power management and communication of data.

Medical contacts lenses preferably are determined applications of Internet of things in a healthcare perspective [16]. As in these technologies has a great future, so far, these organizations have managed to live up to expectations:

In 2012, Microsoft announced the first their own Smart contact lens research, with the ambitions of creating a contact lens which can keep track of diabetic's glucose level with the help of user's tears, which would avoid the users from 'finger-pick blood glucose tests'.

In 2014, Google Life Science (GLS) as partnership with Alcon announced that they could provide a warning system for the users to inform them about the blood glucose level has fluctuated up or down beyond a certain margin.

In 2010, Sensimed a Swiss company developed a non-invasive smart contact lens which was named 'Triggerfish', which would identify the changes in eye dimensions which caused glaucoma. Trigger fish is now FDA-approved and CE-marked hence it is available for marketing and sale in western countries.

6.5.1 Structure of the Smart Contact Lens

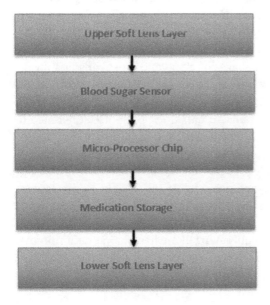

Fig. 4: Structure of Smart Contact Lens.

As shown in the Fig. 4, smart Contact lens has the following components:
- **Upper soft lens layer:** This is the external part of the contact lens designed to remove the components from the user's vision point of view and to protect the components of the lens.
- **Blood Glucose sensor:** This is the essential component of the contact lens; it senses the difference in blood glucose level in contact with users' tears and then sends the signal to the microprocessor chip.
- **Microprocessor chip:** This controls the functioning of the contact lens. Once the data is being received from the sensor it measures the glucose amount and if the rate is above the chip emits a signal for medication.
- **Data Coil:** The use of this coil is to transmit signals sent from the microprocessor chip and also to carry power across the whole components of the contact lens.
- **Medication Storage:** This is a 10 micro storage compartments basically used for medication. Once the signal is emitted to this component, metallic cover of one of the compartments dissolves and medicine is administered to the eye.

6.6 Conclusion

This chapter presents how IoT is integrated in the healthcare domain to enable easy movement for physically challenged persons with less dependency on others. IoT enabled wheelchair with voice controlling or assisting system helps paralyzed people to do an action using just their voice. By using smart wearable devices and other health monitoring equipment embedded with IoT, various health parameters of the patient such as Blood Pressure, calorie count, pulse rate, body temperature, vibration of the body etc., can be remotely monitored by the physicians in real-time. Further these data are stored in databases or cloud and processing or analysis can be done using Machine Learning algorithms and Artificial Intelligence techniques which helps physicians to identify the best treatment for patients and reach the expected outcomes. Doctors or physicians can track patients' adherence to treatment plans or any need for immediate medical attention. And also the patients sitting from their home can take health advice from specialist doctors. This has a major impact on elder patients or people living alone and their families.

A Smart Healthcare Monitoring System using IoT helps to send alert messages and health parameters to the concerned doctors or family members on any disturbances or changes in patients' routine activities. IoT enabled devices are used for tracking real time location of medical equipment like wheelchairs, defibrillators, nebulizers, oxygen pumps and other monitoring equipment. Using IoT a cost effective system can be developed to help blind, dumb and deaf people.

References

[1] AKM Bahalul Haque., Shawan Shurid., Afsana Tasnim Juha. (2020). A Novel Design of Gesture and Voice Controlled Solar-Powered Smart Wheel Chair with Obstacle Detection, IEEE International Conference on Informatics, IoT, and Enabling Technologies (ICIoT), INSPEC Accession Number: 19594447.

[2] Anirvin Sharma., Tanupriya Choudhury., Praveen Kumar. (2018). Health Monitoring & Management using IoT devices in a Cloud Based Framework, IEEE, INSPEC Accession Number: 18070867.

[3] Carla Gómez-Carrasquilla., Karol Quirós-Espinoza., Arys Carrasquilla-Batista. (2020). Wheelchair control through eye blinking and IoT platform, IEEE 11th Latin American Symposium on Circuits & Systems (LASCAS), INSPEC Accession Number: 19534107.

[4] Disha Amrutlal Gandhi ., Munmun Ghosal., (2018). Intelligent Healthcare Using IoT:A Extensive Survey, IEEE 978-1-5386-1974-2.

[5] Islam MM., Sheikh Sadi M., Zamli KM., Ahmed MM. (2019). Developing walking assistants for visually impaired people, a review. IEEE Sens J.

[6] Jafri R, Campos RL., Ali SA., Arabnia HR. (2018). Visual and infrared sensor data-based obstacle detection for the visually impaired using the Google project tango tablet development kit and the unity engine, IEEE Access; 6:443–54.

[7]   Jihun Park., Joohee Kim., So-Yun Kim. (2017). Soft, smart contact lenses with integrations of wireless circuits, glucose sensors, and displays, American Association for the Advancement of Science (AAAS), ISSN: 2375 2548.

[8]   Karmel A, Anushka Sharma., Muktak Pandya., Diksha Garg. (2019). IoT based Assistive Device for Deaf, Dumb and Blind People, Elsevier, International Conference on Recent Trends in Advanced Computing (IICRTAC).

[9]   Mahmud S et al. (2019). A multi-modal human machine interface for controlling a smart wheelchair. 2019 IEEE 7th conference on systems, process and control (ICSPC), Melaka, Malaysia, pp. 10–13.

[10] Md. Milon Islam., Ashikur Rahaman & Md. Rashedul Islam. (2020). Development of Smart Healthcare Monitoring System in IoT Environment Springer Nature Singapore Pte Ltd.

[11] Mohamed Elhoseny., Gustavo Ramírez-González., Osama M. Abu-Elnasr. (2018). Secure Medical Data Transmission Model for IoT-Based Healthcare Systems, IEEE Access (*Volume: 6*) ISSN 2169–3536.

[12] Samruddhi Deshpande. (2017). Real Time Text Detection and Recognition on Hand Held Objects to Assist Blind People, IEEE Conference, 2017.

[13] Sankari Subbiah., S Ramya., G Parvathy Krishna, Senthil Nayagam. (2019). Smart Cane for Visually Impaired Based On IoT, IEEE Xplore, INSPEC Accession Number: 18972324.

[14] Suvarna Nandya.l, Shireen Kausar., (2019). Raspberry-Pi Based Assistive Communication System for Deaf, Dumb and Blind Person, International Journal of Innovative Technology and Exploring Engineering (IJITEE) ISSN: 2278–3075, *Volume-8* Issue-10.

[15] Taiki Takamatsu., Yin Sijie Fang., Shujie Liu Xiaohan., Takeo Miyake. (2019). Multifunctional High-Power Sources for Smart Contact Lenses, IEEE Biomedical Circuits and Systems Conference.

Yogish H K, M Niranjanamurthy, Abhishek K L

Smart Farming Solution using Internet of Things for Rural Area

Abstract: With growing populace throughout the world, farming and production of food progressively profitable and prepared to do exceptional returns in constrained time. The scope for guide experimentation, viability evaluation thru trial and blunders and many others are now not feasible. As per the UN Food and Agriculture Organization, "the world should deliver 70% more food in 2050 than it did in 2006". To satisfy this need, most of the agricultural companies and farmers ought to push the innovation limits of their present practices. The Internet of Things in agriculture guarantees formerly unavailable efficiency, cost and resources reduction, data-driven processes and automation. IoT applications in smart farming for Crop Monitoring, monitoring Climate conditions, Soil quality check, irrigation, agility, Green House automation, Precision farming, Drones in agriculture have been discussed in this chapter.

Keywords: smart farming, drones, precision farming, monitoring, irrigation

7.1 Introduction

Smart Farming is a methodology to improvise the farming technique to improve the efficiency in farming, smart farming helps the farmer to do farming cost efficient and time efficient. Farming is major occupation in India, as whole population of India relies on farming for food which is necessary for survival.

Smart Farming reduces the manual work of farmers and farmers can keep record of the farm in their finger tips on their own cell phones. It helps the farmers from crop wastage and time wastage. Wireless sensors are used to supervise crop conditions [1].

Smart farming reduces the extra use of water on crops as it keeps track of the moisture in soil and weather conditions.

Yogish H K, Department of ISE, M S Ramaiah Institute of Technology, Bangalore, India,
e-mail: yogishhk@gmail.com
M Niranjanamurthy, Department of MCA, Ramaiah Institute of Technology, Bangalore, India,
e-mail: niruhsd@gmail.com
Abhishek K L, Department of MCA, Ramaiah Institute of Technology, Bangalore,
e-mail: abhishekmelkote@gmail.com

https://doi.org/10.1515/9783110725490-007

7.1.1 Benefits of Smart Farming

Following are the benefits of smart farming for real world applications:
- Increased production
- Optimized planning and efficient treatment of crops can increase production.
- Lowered operation costs
- Keeping track of the yield, properly treating them in regular intervals and harvesting reduces human mistakes, increases production and reduces the cost of expenditure.
- Increase in the quality of production.
- Automatically when required amount of water and accurate pesticides are invested on crop it increases the yield hence increases the production rate.
- Accurate Farm and Field Evolution.
- When keeping track of the yield becomes easier, we can easily predict the future crop growth rate and the profit we will earn from the same.
- Improved Livestock Farming.
- Tracking location can improve monitoring and management of livestock.
- Reduced Environmental Footprint.
- All precautions when taken properly like sufficient water and production rate it effect the environment in positive way.
- Remote Monitoring
- Multiple fields can be monitored at the same time.
- Equipment Monitoring
- Farmer equipment's can be properly maintained and monitored in regular intervals based on its production rate.

7.1.2 Drones for Agriculture

Drones are the unmanned aerial vehicle which is an aircraft without a human pilot, in other words, drones are flinging devices remotely controlled either by a remote controller or even by phone. The first drones were enforced before the presence of IoT devices; they were utilized more frequently in military purposes [2]. Drones which are used in the military are called as combat drones; these drones are fast and robust, long-lasting or jet engine steam-powered. The civilian drones are completely different from the drones which are used in the military. Likewise, civilian drones are relatively slow, lightweight and fragile.

Drones are used in many ways and for various types of activities, Agriculture is one among them where drones are used nowadays for a variety of uses from the soil and crop analysis to planting and spraying pesticides, also drones can identify which regions of the field are drier and measures are taken to irrigate that area of the field, drones can also monitor crop health and fungal infection also drones help analyze the soil as shown in the Fig. 1.

Fig. 1: IoT in Agtech: Drones in digital farming.

Drones are flying over the sector and take high-resolution images using a camera or a sensor [7], the pictures are captured from different bands and spectrums, images captured by the drones are called as raw data which will be analyzed in various ways to make agriculture better.

One of the big challenges for farmers is to deal with bad weather conditions; drones can also monitor the weather so that farmers can take precautions before the bad weather hits. Drones can do many other things like analyzing soil condition, planting crops, fighting against infection and pests, agricultural spraying, surveillance of the corps etc. Farmers can obtain various types of information from the drones like data collected by the drones gets translated into useful and comprehensive information for the farmers like,

- Height and density of the plants.
- Plant counting, plant size, plot statics.
- Infestations, phrenology, leaf area, water needs, anomaly detection.
- Field and Soil analysis drones produces accurate 3D maps for analysis of soil which is useful in seed planting.
- Spraying pesticides to crop, drones can test the floor and spray the right measure of liquid.
- Health assessment drones can identify which part of the farm or land is in good condition.
- In irrigation, dry part of the field can be identified using drones which have thermal or spectral sensors.

7.1.3 Precision Farming

Agriculture is one of the most important and also a useful field in this world. Day by day, agriculture has been using technology as a common element. So many applications and scientific tools have been invented to make work much easier. For developing Internet of Things (IoT) in precision farming, a better understanding of any particular field is very important. We can say that the Internet of Things is a solution for any problem related to communication in agriculture. Many problems faced by Precision Farming as were all in the solution process by many Agriculture Intelligence. According to modern farming, good communication between farmer, plants, trees, animals, machines is very important. Adoption of IoT normally requires large scale farming and appropriate large agriculture and should have a secured communication [11].

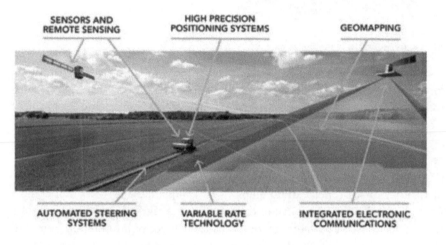

Fig. 2: Precision Farming in India.

Large farmland with a good type of soil having perfect temperature for yield and also necessary nutrients for better harvest can be done with the help of IoT. This affects the productivity of agricultural land. By using these kinds of technology in agriculture, we can see rapid growth in modern science and IoT [5]. This also reduces the necessity of unwanted sources like the axe and other tolls for agriculture but helps in improving the better-modernized way of distributing the pesticides, fertilizer, or water.

Mainly, Precision farming as shown in Fig. 2, aims at providing a platform to enhance the production of agriculture and at the same time, optimizing the cost of the farmers as well as preserve the sources in a better systematic way which will be a better practice on the agriculture. Precision farming also aims at the generation of data through sensors and also on analyzing of the data for evaluating the required action to make a good decision about the crops. IoT can be used in smart farming in many different successful ways. It especially aims in areas such as field observation, storage monitoring, livestock monitoring and farm vehicle tracking. Even to prevent pesticides and insects from affecting the yield, sensors can be used to do detect them and to take certain actions on it and also maintaining them in a systematic way [9]. Even farmers can grow and learn more about IoT devices and improve their day-by-day life in many useful ways. Animals which are used in field work can also be reduced with the help of modernized IoT technology.

These IoT devices can also find out which animal is infected with any problems or disease and which animal disease will infect the whole group of animals; hence every development has to be made on this case and should be updated to the owner.

7.1.4 Crop Monitoring

Forming in India is done using multiple ways. Most of the farmers have lack of knowledge, which makes them do agricultural activities based on predictions, makes it even more erratic. IoT permit the objects to be recognized and modulate to an extent across prevail network model. The sensors sense the field specifications such as temperature, humidity, moisture, and fertility in the farm. The recorded values are authenticated and later send to the WI-FI module and from using cloud network the validated data is sent the farmers mobile or laptop from the Wi-Fi module [15]. If there is any necessity of urgent care to the field farmers are notified by the SMS to their mobile.

To keep track of the water quantity a design is initiated using threshold value of temperature, humidity, moisture and fertility. The Farmer can operate the motor from anywhere which is at fingertip.

7.1.4.1 Features of Crop Monitoring are

1. SMS notifications.
2. Data collection and make decisions.
3. Monitoring and operate.
4. Profitable.
5. Feasible.

Tools used: Raspberry pi Micro controller, Power supply, Pi camera.

Sensors used: Moisture, Temperature, Humidity Sensor and Water Level Sensor.

Various sensors are implemented to collect the information as shown in the Fig. 3. This system is designed in smart Agriculture in order to observe the farm management. The system is built using Micro controller of Raspberry-pi, Humidity sensor, Water level sensors Moisture and Temperature sensor along with LCD Display. In the first instance built system collects all the guideline concerning crop and exhibit in the LCD-Display. Using the assistance of Pi camera, the system will anticipate and anatomy for animal detection and the same is conveyed to the server [3] .

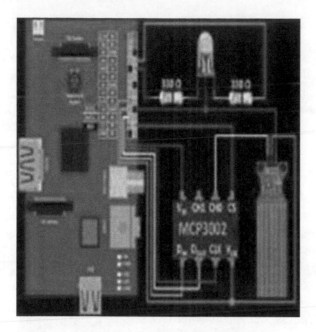

Fig. 3: Connection diagram for Crop Monitoring.

7.1.5 Weather Monitoring

Weather condition simulates a vital role in our everyday life. Weather monitoring facilitate in various ways. The climatic state is noticed to keep up the good development in crops and to make sure the safe functioning habitat in industries, etc. The sensors are the compressed electronic devices used to measure the physical and environmental parameters. Sensors can be used to observe the weather conditions, that outcome will be precise and the complete system will be brisk and low power consuming [13].

7.1.5.1 Benefits

1. Enhanced for monitoring & controlling of atmosphere conditions.
2. Automatic message generation & transmission
3. It does not obligate any human inadvertence.
4. Anterior alert of weather conditions.
5. Profitable
6. High precision.
7. Pure Intuition.

7.1.5.2 Sample System Design for Weather Monitoring using IoT

System is built with Raspberry-pi which envision and store distinct climatic specification with the assist of sensors connected to Raspberry gets all the data. Then on the output aspect, LCD is equated meant for screening the result and OFF, ON broadcast intended for connection of the server. Additionally, 5 V, 1 A power is supplied to the Raspberry Pi through USB slot as shown in Fig. 4.

An SD card of 8 GB is cast-off to reserve the operating system in addition to all programs and files required. Using USB port keyboard and mouse are attached to Raspberry Pie. Monitor is associated to the Raspberry Pi board via HDMI port with the help of HDMI to VGA cable. Various weather monitoring sensors such as temperature sensor, light sensor, humidity sensor and rain sensor are connected to the GPIO pins of raspberry pi. Ethernet port is used to give internet connection to the system via LAN. By providing username and password user can connect to the server to predict current weather condition of remote location. After the user logs in the processed data which is stored in raspberry pie using the cloud server is generated in the form of the graph which is in hourly and daily predicted data.

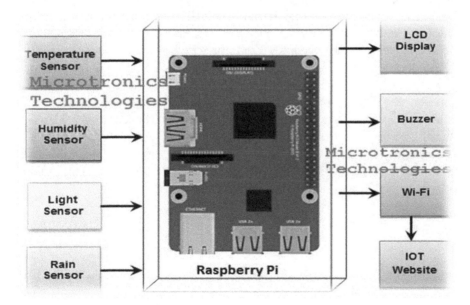

Fig. 4: Connection diagram for Weather Monitoring.

The weather forecasting does essential role in agriculture. It helps at places like volcano and rain forests. It is strenuous for a human being to rest for extended duration at corresponding location.

7.1.6 Green House Monitoring

Greenhouse cybernetics System is the practical submission wherein the farmers within side the backwoods might be profited via way of means of automatic tracking and direct of greenhouse environment. System restores the face-to-face administration of individual. Greenhouse is structures where plants are grow up in controlled fashion. Because of development and insufficient land accessibility it is disparity to set up the Greenhouses which will be restrained for a part of agriculture. With the evolution of automation, we can empower and prefect the various Greenhouses using IoT by the mid locality [14].

Greenhouse farming is an approach which intensifies the flex of crops, vegetables, fruits etc. Greenhouses manage environmental edge in two ways; either through handbook interference or a comparative control appliance. However, since handbook interference has drawback of energy and production loss, labor cost will be high, hence these methods are insufficient [12]. This IoT enabled smart greenhouse system not only monitor smart but also manages the climate. Thereby eliminates need of human interference.

Making use of IOT technology the farmer not only gather data from GREEN HOUSE but also farmer can manage it over 'INTERNET', this technology has ability over largest distance, simple access, fast data processing and billions of applications can get associated over web. IOT can initiate farmer to associate with a existing digital technology to build up profit from farmers and abundance of crop, where we require technology for growth of any country considering food yields is a basic need of any country, and this kind technology we must use in basic need, so trying to implement this tremendous technology with a basic need [8].

According to plant necessity various sensors such as temperature sensor, light sensor, humidity sensor and rain sensor, compute the environmental specifications for regulating the atmosphere in smart greenhouse. Then for remote accessing of the system, a cloud server is created by connecting IoT.

7.1.6.1 Benefits

1. **Sustain best Micro-Climate environment:** Sensors admit farmers to collect varied information at exceptional granularity. Using decisive climatic aspect involving, temperature sensor, light sensor, humidity exposure and carbon dioxide, instantaneous data across the greenhouse can be collected. While navigating energy productivity the data provokes appropriate alignment to HVAC and illumination settings to uphold the leading circumstances for plant precision.
2. **Improved Irrigation and Fertilization practice:** along with surrounding circumstances, IoT enabled greenhouses allow farmers to up-to-date the crop

condition which guarantees irrigation and fertilization recreation with the practical requirements of urbane plants designed for maximized yields.

3. **Infection Control and Disease Outbreak Avoidance:** Crop disease is preserved farming summons, through every eruption taking so much tax on crop margins. With the help of provided information the farmers can provide proper treatment to the plants.

4. **Thefts Prevention and Improved Security:** Greenhouses with admired crops are an in danger of extinction plan meant for thieves. IoT sensors in smart greenhouses contribute a modest framework to monitor door stature and determine unsure venture.

7.1.7 Smart Irrigation System

Using Mobile devices, a device functions can be monitored using IoT technology. The IoT is concerned with interdependent transferring recipients that are initiated at various positions that are perhaps far from each other. At the present time water shortage is a biggest distress for agriculture. The preferred system has been plotted to control excess water flood into the farming land. The information about the crop is collected from Temperature sensor, humidity sensor, moisture sensors and these values are sent to the authorized Internet Protocol address [7]. Android applications are used to collects the data constantly from that assigned Internet Protocol address. If the moisture content level goes below or above the particular limit, Arduino micro controller manages automation of irrigation system [4].

7.1.7.1 Features of Smart Irrigation System are

- Increases production rate.
- Low water exploitation.
- No manpower required.
- Require smaller water source.
- Reduce soil erosion and nutrient leaching.

Tools used:
- Temperature and humidity sensor.
- Soil moisture sensor.

7.1.7.2 IoT based Sample System for Smart Irrigation

In Arduino based automated system for smart irrigation, three sensors are attached to the controller and observed values from sensing devices are transferred to mobile applications. As shown in the Fig. 5, soil moisture sensor, humidity and temperature sensor, rain sensor are attached to Arduino board. Wireless sensor network of the system is used for simultaneous sensing of data from various sensors. This system furnishes uniform and vital level of water for the agricultural farm.

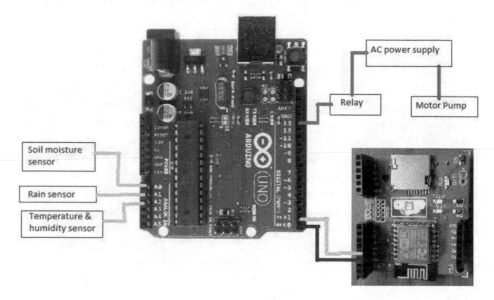

Fig. 5: IoT based Smart Irrigation System.

7.1.8 Conclusion

This chapter presents how IoT is integrated in Smart farming to enable farmers to adopt IoT technologies to produce better quality crops. Drones are useful in analyzing soil condition, planting crops, fighting against infection and pests, agricultural spraying, surveillance of the corps etc. The farmer can monitor crops condition and automatically sprinkler the water to the crops at his finger tip using IoT. Smart Irrigation using IoT, increases production rate, low water exploitation, reduces manpower and Reduce soil erosion and nutrient leaching.

References

[1] S. R. Nandurkar, V. R. Thool, R. C. Thool, "Structure and Development of Precision Agriculture System Using Wireless Sensor Network", IEEE International Conference on Automation, Control, Energy and Systems (ACES), 2014.

[2] Joaquin Gutierrez, Juan Francisco Villa-Medina, Alejandra NietoGaribay, and Miguel Ángel Porta-Gándara, "Robotized Irrigation System Using a Wireless Sensor Network and GPRS Module", IEEE transactions on instrumentation and measurement.

[3] Dr.V Vidya Devi, G. Meena Kumari, "Constant Automation and Monitoring System for Modernized Agriculture", International Journal of Review and Research in Applied Sciences and Engineering (IJRRASE) *Vol3* No.1. PP 7–12, 2013.

[4] Y. Kim, R. Evans and W. Iversen, "Remote Sensing and Control of an Irrigation System Using a Distributed Wireless Sensor Network", IEEE Transactions on Instrumentation and Measurement, pp. 1379–1387, 2008.

[5] Q. Wang, A. Terzis and A. Szalay, "A Novel Soil Measuring Wireless Sensor Network", IEEE Transactions on Instrumentation and Measurement, pp.412–415, 2010.

[6] Arampatzis, T.; Lygeros, J.; Manesis, S. An overview of utilizations of remote sensors and Wireless Sensor Networks. In 2005 IEEE Worldwide Symposium on Intelligent Control and thirteenth Mediterranean Conference on Control and Automation. Limassol, Cyprus, 2005, 1–2, 719–724.

[7] Orazio Mirabella and Michele Brischetto, 2011. "A Hybrid Wired/Wireless Networking Infrastructure for Greenhouse The board", IEEE exchanges on instrumentation and estimation, vol. *60*, no. 2, pp 398–407.

[8] N. Kotamaki and S. Thessler and J. Koskiaho and A. O. Hannukkala and H. Huitu and T. Huttula and J. Havento and M. Jarvenpaa (2009). "Remote in-situ sensor organizes for horticulture what's more, water observing on a stream bowl scale in Southern Finland: assessment from an information clients point of view". Sensors *4*, 9: 2862–2883. Doi: 10.3390/s90402862 2009.

[9] Baker, N. ZigBee and bluetooth – Strengths and shortcomings for modern applications. Compute. Control. Eng. 2005, *16*, 20–25.

[10] Dr. M. Newlin Rajkumar, S. Abinaya, Dr. V. Venkatesa Kumar "Intelligent irrigation system – an iot based approach" IEEE International Conference on Innovations in Green Energy and Healthcare Technologies (ICIGEHT'17), 978-1-5090-5778-8/17/$31.00©2017 IEEE.

[11] Shweta Bhatia, Sweety Patel, "Analysis on different Data Mining Techniques and algorithms used in IOT", ISSN: 2248-9622, Vol. *5*, Issue 11, (Part – 1) November 2015, pp.82–85.

[12] Nguyen Cong Luong, Dinh Thai Hoang, Ping Wang, Dusit Niyato, Dong In Kim, and Zhu Han "Data Collection and Wireless Communication in Internet of Things (IoT) Using Economic Analysis and Pricing Models: A Survey" arXiv:1608.03475v1 [cs.GT] 11 Aug 2016.

[13] Prof. K. A. Patil, Prof. N. R. Kale "A Model for Smart Agriculture Using IoT" IEEE | December 2016.

Vaishnavi A S, Dr. Sumana

Assistance Device for Visually Impaired based on Image Detection and Classification using DCNN

Abstract: Based on the statistics World Health Organization (WHO) has provided, there are aroundeighty five million visually disabled people worldwide. Among them thirty-nine million people are completely blind. The major difficulties faced by them are obstacle detection and this can be eradicated using Deep Learning. Deep Learning provides computer vision to the system which makes decisions based on training algorithms. The main goal of the proposed work is to present a method for detecting the objects in their surrounding or in a given environment. The proposed work presents a cost-effective approach for detecting the objects via images in a given environment. This approach detects presence of certain objects in the scene, regardless of their position using multi-labelling. Our proposed solution is based on Deep Learning Convolution Neural Network which makes use of LeakyRelu activation function along with Batch Normalization layer and output layer activation functions are Linear functions. Suppose if we place the camera at Vision impaired person chest, then we can calculate the distance between that particular object and the person which is through USB camera and USB laser which is integrated with the odroid system and generates an audio output. This approach is used for detecting certain objects in a given frame that is in his surroundings and calculates the distance between Visually Impaired person and the object and then provides an audio based output.

Keywords: deep learning, convolution neural network, visually impaired, image multi labelling, image detection, object detection

8.1 Introduction

For visually impaired or blind people, the important aspect to be considered is their mobility or navigation around their surroundings. For the normal people, there are maps for travelling in an unknown place. But for visually impaired people, it is very

Vaishnavi A S, Dept. of Information Science and Engineering, M S Ramaiah Institute of Technology, Bangalore, India, e-mail: vshnvgowda@gmail.com

Dr. Sumana, Dept. of Information Science and Engineering, M S Ramaiah Institute of Technology, Bangalore, India, e-mail: sumana.a.a@gmail.com

https://doi.org/10.1515/9783110725490-008

difficult for them to move each step. In this era, where technology has evolved so much, it is very essential to develop a model to at least full-fill their basic needs at least like their navigation around the surroundings to make their life easier a bit. One way is to use stick for object detection but for navigation they need to completely rely on someone to help them to with the surrounding details or if they're in a new area or unfamiliar area, they can't use the maps, they need someone to help [2]. With sticks they can only detect the objects on their way while memorising the path. Visually impaired people, blindly rely on other factors such as touch of the objects or audio based signals as well [21].

To make visually impaired people life easier, the proposed article presents a model, which not only go with stick that is object detection along with alarming the user about the surrounding objects with the distance between the user and the surrounding objects and suggesting the direction of the path to move. So just with the help of traditional ways that is using sticks, they cannot understand the perception within that scene [1]. However, information about the presence of objects near them will be available through sticks, but no information on the type of object which blocking their way. In many cases it is very important to know about the object which is blocking their way, for example door or stair case or trees or any vehicles. It is very difficult for them to open the door or climb the stairs without any external aid.

All visual information that is written on boards in public places which show the direction for navigation are very difficult to analyse or get that information for blind or visually impaired people [4].

When it comes to computer vision object detection and classification are the trending fields. With the study of object detection and its classification, there is an inclination in the curve for their applications. Computer vision is another one aspect which is broadly used in building artificial intelligence application. Object detection and its classification are the building blocks for computer vision which has different machine learning and deep learning algorithms incorporated. For any model considered network is trained using training examples/data set and on the basis of this training, the model will predict when an unseen data is presented during testing and accuracy can be predicted when the model identifies the required data with less error rate. Deep learning and machine learning are the subset of artificial intelligence [14].

Deep learning acts as human brain while machine learning behaves as neural network of the model. Auto assistance system is used to identify the objects within a given surrounding, that is not outdoor environment. Then classify the detected objects, since outdoor is not considered, the objects detected might be person, chair, door, table or stairs and so on. Convolution Neural Network makes use of Multi-Layer Feed-forward neural network since the proposed methodology proposes the model based on deep learning. When considered fully connected neural networks, weights are shared among each neuron. You look only once Yolo is used to improve the accuracy by identifying the object then classifying the object along with the location of the object,that is object localization.

The proposed article presents a model that is assistance device for helping visually impaired person based on image detection and classification to make visually impaired person life little easier [8]. This model makes use of odriod system, why this system is because it can be integrated with USB cam as well laser for identifying the objects in the given environment and the distance of them from the user. So this system can be put or pinned in the chest. The system will detect the objects within the given environment and gives the audio based output. Recent trends are based on the utilization of computer framework to aid the navigation and the solution for it.

Consider an example where smart phones help to move from one place to another both inside and outside environment. The proposed article presents the model which makes use of computer vision for capturing the images and detecting the obstacles or object and then classifying them. Convolutional neural network detects the objects within the frame and computer vision provides vision to the system and works based on the training dataset used. For image sharpening or removing the outliers or getting the required features region based convolutional neural network. This helps to detect the object present in the frame with different position [3].

Deep Convolutional Neural networks is a leading technology in object detection, image classification. Since deep convolutional neural networks uses relu activation function, this allowed faster training. If the network is very complex, then more than one GPU's can be used this is similar to multi column deep convolution neural network. Errors can be reduced by overlapping max pooling layers. At specific layers, GPU interaction can be done. Overfitting is reduced by pooling that is data augmentation. To obtain maximum values of the image portion covered by the kernel, max pooling is performed.

Max pooling reduces the noise in the input along with the dimensionality reduction. Convolutional neural network will have these layers increased if the input dataset contains more complex images or to capture more low level detailed information of the image. The output of the final layer can be fed as input for other neural networks, flatten the image which is transposing the row to column vector. Then backpropagation algorithm can be applied to get the network for each training iteration. Then after several series of epochs, model will learn through experience and would provide very accurate results.

8.1.1 Motivation

Proposed methodology presents a model that is assistance device for helping visually impaired person based on image detection and classification to make visually impaired person life little easier. This model makes use of odriod system, why this system is because it can be integrated with USB cam as well laser for identifying the objects in the given environment and the distance of them from the user. So this system can be put or pinned in the chest. The system will detect the objects within the given environment and gives the audio based output.

8.1.2 Scope

Recent trends are based on the utilization of computer framework to aid the navigation and the solution for it. Consider an example where smart phones help to move from one place to another both inside and outside environment. Theproposed article presents the model which makes use of computer vision for capturing the images and detecting the obstacles or object and then classifying them. Convolutional neural network detects the objects within the frame and computer vision provides vision to the system and works based on the training dataset used. For image sharpening or removing the outliers or getting the required features region based convolutional neural network. This helps to detect the object present in the frame with different position.

8.1.3 Problem Statement

Proposed methodology presents the model which makes use of computer vision for capturing the images and detecting the obstacles or object and then classifying them. The proposed article presents a model that is assistance device for helping visually impaired person based on image detection and classification to make visually impaired person life little easier. This model makes use of odriod system, why this system is because it can be integrated with USB cam as well laser for identifying the objects in the given environment and the distance of them from the user. So this system can be put or pinned in the chest. The system will detect the objects within the given environment and gives the audio based output.

8.1.4 Objectives

Proposed methodology presents a model that is assistance device for helping visually impaired person based on image detection and classification to make visually impaired person life little easier. This model makes use of odriod system, why this system is because it can be integrated with USB camera as well laser for identifying the objects in the given environment and the distance of them from the user. The proposed research work presents the model which makes use of computer vision for capturing the images and detecting the obstacles or object and then classifying them. Convolutional neural network detects the objects within the frame and computer vision provides vision to the system and works based on the training dataset used. For image sharpening or removing the outliers or getting the required features region based convolutional neural network. This helps to detect the object present in the frame with different position.

8.2 Existing Functionality

Existing face recognition systems used Haar Cascade algorithms which uses Haar features and this can also be used for object detection. Haar Cascade has lower accuracy or performance metrics when compared to Convolutional Neural Network and other machine or deep learning algorithms which are used for identifying the objects [1]. Haar classifier can also be used for detecting or tracking people in the CCTV or any live video streaming. This can be achieved using the functions of the classifier that is Haar classifier which results in high speed calculation. This technique can be used for training the model for unseen data. Tracking the people in live video can be achieved through sampling techniques along with resampling and this sampling helps in predicting the position of the person in the video with high accuracy.LCuimei et.al in [3] suggested three different Haar classifiers for face recognition like consider one classifier for skin hue histogram matching and weak classifier for eye and lips detection.

The proposed article presents different approaches which can be used to detect multiple objects simultaneously. Dataset used to train the model was coco dataset. On CPU, Haar classifier is faster but convolutional neural network is much faster and accurate when compared to haar. Convolutional neural network detects multiple objects simultaneously for real time applications. But on GPU training is faster, when compared to CPU and processing is also extremely faster which reduces high computational complexity [2].

The proposed work also presents the drawbacks of Viola Jones classifier and this gave high accuracy with ideal rejection rate. But when compared to Haar classifier, CNN has is cost effective and computation overhead is very less when used along with several other methods or algorithms with constant learning rate and object detection parameters. The proposed article presents the Shallow network which verifies the high accuracy for good detection [3].

Computer vision is nothing but extracting the useful features analysing and understanding the required information for the single image or the set of images given. But using convolutional neural network, output will be having high level features from top layer of the network. However, these features may not contain the information in detail. Therefore, the output of one just layer might not contain most of the information required, so there needs to be a way for fusion of multiple layers. Drawback of this proposed methodology is trying to train the model on CPU, since it requires high memory management and computational complexity [4].

CNN architecture is very simple and detection is limited [4] so to increase the efficiency depth network is also combined [14] along with convolutional neural network. Bag of visual words model is also one of the commonly used algorithms for image classification since it has features for detecting the unique features of an image. When trained this with new dataset containing 6 classes then when it is compared with CNN which would take dataset containing 10 classes for the same, showed less accuracy [5].

The proposed article presents a model which can be used to improve the performance of the convolutional neural network used for object detection. Several different object classes were considered, 5 layers of convolutional layers were applied, then compared the proposed model with different bag of words approaches. The accuracy was around 90.12%, which was better than bag of words. Drawback of this approach was this model cannot be used for real time application where images are captured and processed the data [6].

All Artificial intelligence model requires big data for training. So when considered, object detection requires a lot of images for training. When dataset downloaded from web, it must surpass this data pre-processing technique. This proposed methodology presents the pre-processing techniques used for training YOLO [7]. Drawback of this proposed methodology is all training images should be of same size. But yet the output of the model has the image which is of same size as the input image, almost similar.

Haar Cascade has lower accuracy or performance metrics when compared to Convolutional Neural Network and other machine or deep learning algorithms which are used for identifying the objects [8]. Haar classifier can also be used for detecting or tracking people in the CCTV or any live video streaming. This can be achieved using the functions of the classifier that is Haar classifier which results in high speed calculation. This technique can be used for training the model for unseen data.

Tracking the people in live video can be achieved through sampling techniques along with resampling and this sampling helps in predicting the position of the person in the video with high accuracy.L Cuimei et.al in [3] suggested three different Haar classifiers for face recognition like consider one classifier for skin hue histogram matching and weak classifier for eye and lips detection. The proposed methodology presents different approaches which can be used to detect multiple objects simultaneously [9].

There are two types of classification in neural networks, single and multi-layer classification. Single layer classification is nothing but, only one output neuron. But with multi class classification we can have output units same as input units. Soft-max activation function works by assigning the probability for the outcomes generated [3]. Then output or data is classified into class which has highest probability value. But for multi-labelling, final outcome should be independent of each other so softmax is wrong choice here. This is why sigmoid activation function is used for the final layer. The output or score values will between 0 and 1 and independent [10].

Yolo is used is along with the pre-processed images which are used for training, for training the model. Whenever any model is designed two important things needs to considered is [11],

1. The size or dimension of the output image when compared with input image [3].
2. The object dimension area occupied in the image should be equal in both training and testing dataset or with the detected images [6].

For object detection and their classificationYolo v3 can be used [7], which results in lesser error rate and time taken is very low and detection or computation speed is very high in real environment [10]. odriod is a combination of open and android. This board can be used not only for android, but for Linux distributions as well. This board has USB charging serial port and the hardware is quite secured, that is accessible only by the parent company. It is very small in size like wearable device and obviously it is a portable device. The proposed methodology presents Odroid with XU4 version. This board is considered because it is cost affection and software can be updated very easily and user friendly.

For object detection, Pascal VOC datasets are considered which had at least 1k images and 20 labels were considered and different object identification and classification algorithms were considered and with recurrent convolutional neural network, the error rate was very low [8].

The proposed methodology presents a model that is assistance device for helping visually impaired person based on image detection and classification to make visually impaired person life little easier [8]. This model makes use of odriod system, why this system is because it can be integrated with USB cam as well laser for identifying the objects in the given environment and the distance of them from the user. So this system can be put or pinned in the chest.

The system will detect the objects within the given environment and gives the audio based output. Recent trends are based on the utilization of computer framework to aid the navigation and the solution for it. Consider an example where smart phones help to move from one place to another both inside and outside environment.

The proposed article presents the model which makes use of computer vision for capturing the images and detecting the obstacles or object and then classifying them. Convolutional neural network detects the objects within the frame and computer vision provides vision to the system and works based on the training dataset used. For image sharpening or removing the outliers or getting the required features region based convolutional neural network. This helps to detect the object present in the frame with different position [3].

Artificial Neural Network was compared against Jones on how accurately the moving objects are detected and classified. Usually ANN identifies more features for classification detects multiple objects which are moving [9]. Drawback are some classes in the dataset occurs more frequently, for example if a dataset contains 900person image and 100 table images, then if the model is trained on this type of dataset, then model learns to predict person every time, and the error rate will increase and accuracy will be dropped. So one way to overcome such issues are Sampling.

When it comes to computer vision object detection and classification are the trending fields. With the study of object detection and its classification, there is an inclination in the curve for their applications. Computer vision is another one aspect which is broadly used in building artificial intelligence application. Object detection and its classification are the building blocks for computer vision which has different machine learning and deep learning algorithms incorporated [13].

For any model considered network is trained using training examples/data set and on the basis of this training, the model will predict when an unseen data is presented during testing and accuracy can be predicted when the model identifies the required data with less error rate. Deep learning and machine learning are the subset of artificial intelligence [14].

To obtain maximum values of the image portion covered by the kernel, max pooling is performed. Max pooling reduces the noise in the input along with the dimensionality reduction. Convolutional neural network will have these layers increased [15] if the input dataset contains more complex images or to capture more low level detailed information of the image. The output of the final layer can be fed as input for other neural networks, flatten the image which is transposing the row to column vector. Then backpropagation algorithm can be applied to get the network for each training iteration. Then after several series of epochs, model will learn through experience and would provide very accurate results [9].

Deep learning acts as human brain while machine learning behaves as neural network of the model. Auto assistance system is used to identify the objects within a given surrounding, that is not outdoor environment. Then classify the detected objects, since outdoor is not considered, the objects detected might be person, chair, door, table or stairs and so on. Convolution Neural Network makes use of Multi-Layer Feed-forward neural network since the proposed methodology proposes the model based on deep learning. When considered fully connected neural networks, weights are shared among each neuron. You look only once Yolo is used to improve the accuracy by identifying the object then classifying the object along with the location of the object,that is object localization [16].

Commonly used techniques can also include some of the devices which can be worn that give stimulated response about navigation [15] using several vibrating motor. Another technique which can be related to sensing the sound that is with the aid of sensors. Here a scene can be visualised into acoustic or some stimulated display tactile [16]. Amplifying the echo regarding the surrounding can also be used to get the object information [17]. Smart equipments like glasses with ultrasonic signals or waves to identify the obstacles [18].

There are many other disadvantages of these blind aid devices, like some of them may be very invasive for wearing in the ear. Blind people might find many other problems like which may block their tongue or hands as well [19]. So with these devices they will never feel free and they need to hold the device, its weight and handling these devices at public is major challenge. While managing these devices and its load, this will distract from their primary work.

Performance can be calculated using precision and recall. Average value of precision is usually numerical value, which can be used to compare the performance across different algorithms. Based on precision and recall graph curve, weighted mean of each object can be calculated. Based on the ground truth, these performances can be calculated [21]. Haar features or Viola and Jones Haar like features are just like kernel of the convolutional neural network except that in convolutional neural network, weights or the values of the kernel will be altered based on the experience while training but in Haar the values or features are assigned or determined manually.

UsuallyHaar feature based cascade classifier is used for feature detection with limited features since all the features are manually determined. Suppose for Haar algorithms if the input is edge detection of a face and if there are some slight changes in the face then that won't be detected using haar but convolutional neural network detects it well with greater accuracy. Haar based classifier can be used for small datasets, since there is no training required. Accuracy is calculated based on the number of predictions which are correct when divided by number of examples or dataset used.

Usually these devices require alot of training and using these devices is very risky or difficult task when children are considered. Since it is a device, it's cost is also very high when compared to the proposed methodology. When considered these devices, they will not be completely tested or some of them will be in preclinical stage or some will have just the prototype [20].

Fully connected convolutional layer is simply feed forward neural network. These layers would be the layers from last few layers connected in the network. Input to the fully connected neural network will be the output of some convolutional layer or other final pooling layer. That is output of any fully connected or convolutional neural network is a three dimensional matrix. This output will be unrolled to its values in the vector and then fed as input to another feed forward network [22].

Since when the images are captured, there might be several objects in the image, which are semantically related, for example furniture. This is when multi labelling comes into picture, since the same image can belong to multiple other classes. There are several others ways for classification, one way is binary classification, that is m independent classifiers for 'm' classes of data. Then the final layer would consist of m sigmoidal activation function. Then based on the threshold on the threshold of this logistic layer, prediction can be made. But theonly drawback for semantically related features, few labels might have been ignored. But training model is easy [8].

For outdoor navigation, GPS is the best way. But for indoor navigation for the blind people it's very difficult to find the objects in indoor environment. The proposed article presents a model named ARIANNA that is path Recognition for Indoor Assisted Navigation with Augmented perception. This model helps to detect some path which is painted on the floor. And as soon as the path is detected a notification will be made that is by beeping. Then QR Code will be generated. But the distance between the user and object is not calculated and just beeping won't help if the person using is somewhere in the crowd but in indoor environment [16].

For object detection and their classificationYolo (You Look Only once) can be used, which results in lesser error rate and time taken is very low and detection or computation speed is very high in real environment [10]. odriod is a combination of open and android. This board can be used not only for android, but for Linux distributions as well. This board has USB charging serial port and the hardware is quite secured, that is accessible only by the parent company. It is very small in size like wearable device and obviously it is a portable device. The proposed methodology presents Odroid with XU4 version. This board is considered because it is cost effective, software can be updated very easily and is user friendly [12].

The proposed methodology presents the model which makes use of computer vision for capturing the images and detecting the obstacles or object and then classifying them. Convolutional neural network detects the objects within the frame and computer vision provides vision to the system and works based on the training dataset used. For image sharpening or removing the outliers or getting the required features region based convolutional neural network. This helps to detect the object present in the frame with different position [20].

Artificial Intelligence helps to view the world through computer vision as human does. It is nothing but bridging the gap between humans and computers. It helps in analysing the world as humans does and perceiving the inputs in the similar fashion how human brain works. Computer Vision is the domain enables the computer to predict and make decisions after constant analysis, just like human. The important aspect of this field is to view the world as human does, perceive the input and process the information similar to the human brain. This domain is widely used in Image detection or video recognition system, natural language processing, Image classification, recreating the media, movie recommendation systems and so on. Convolutional neural network is the algorithm used in computer vision along with the advancements in deep convolutional neural network [22].

8.3 System Design

8.3.1 Odriod Board

As the name itself suggests, odriod is a combination of open and android. This board can be used not only for android, but for Linux distributions as well. This board has USB charging serial port and the hardware is quite secured, that is accessible only by the parent company. It is very small in size like wearable device and obviously it is a portable device. The proposed work presents Odrold with XU4 version. This board is considered because it is cost effecting and software can be updated very easily and user friendly. This board even have Samsung processors which increase the performance of this board. This board even have large memory which is available to store large amount of images captured in real time. Throughput can be increased with parallel computation, due to OpenCL incorporated. This board has two ports that is USB ports for capturing HD live streaming for two cameras. At dawn times as well, it gives clarity images. Camera can be attached to the chest along with camera.

8.3.2 Deep Convolutional Neural Networks

As shown in the Fig. 1 [21], Deep convolution networks are used for object detection which consists of convolutional layers, pooling layers and fully connected layer for classification. The dropout is considered for max pooling to reduce overfitting. Based on the training examples used n-fold cross validation is used to train and test the model.

In the above diagram output of convolutional operation is known as feature map will be converted into input vectors. Then all these vectors are combined to create a model. These vector signifies the features required. When any input image is fed into the convolutional layer, apply filters or choose proper parameters along with strides or padding required. Perform the operation that is convolutional on the input. For non-linearity apply Relu Activation function to the convolution operation. Pooling is performed to reduce the dimensionality. Until the error rate is reduced convolution layers are added. Then this flattened output is fed to fully connected to convolution layer. Then model will output based on the activation function used and object detection and their classification is done. Activation functions like sigmoid (logistic) or soft-max is used to output as person, or table or door in our model.

Overfitting in Deep Convolutional Neural is reduced by pooling that is data augmentation.To obtain maximum values of the image portion covered by the kernel, max pooling is performed. Max pooling reduces the noise in the input along with the dimensionality reduction. Convolutional neural network will have these layers increased if the input dataset contains more complex images or to capture more low level detailed information of the image. The output of the final layer can be fed as input for other neural networks, flatten the image which is transposing the row to column vector. Then backpropagation algorithm can be applied to get the network for each training iteration. Then after several series of epochs, model will learn through experience and would provide very accurate results.

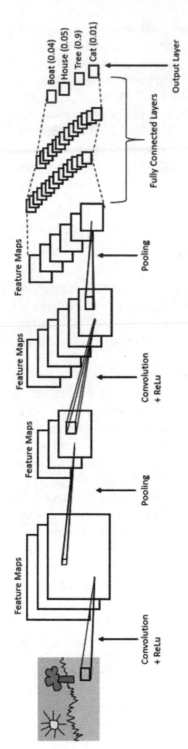

Fig. 1: Deep Convolution Neural Network.

For object detection and their classificationYolo v3 can be used [7], which results in lesser error rate and time taken is very low and detection or computation speed is very high in real environment [10]. odrIod is a combination of open and android. This board can be used not only for android, but for Linux distributions as well. This board has USB charging serial port and the hardware is quite secured, that is accessible only by the parent company. It is very small in size like wearable device and obviously it is a portable device. Theproposed article presents Odroid with XU4 version. This board is considered because it is cost effecting and software can be updated very easily and user friendly.

8.3.3 Convolutional Neural Networks

Artificial Intelligence helps to view the world through computer vision as human does. It is nothing but bridging the gap between humans and computers. It helps in analysing the world as humans does and perceiving the inputs in the similar fashion how human brain works.Convolutional neural network is a deep learning algorithm which can take input assign the weights based on the objects in the image and identify these objects or different objects can be identified. Image pre-processing required is very lower when compared to other algorithms available for classification or recognition of objects. Using convolutional neural network, the model can identify any filters applied in the image without any training and they have the ability to learn these filters or characteristics of the image.

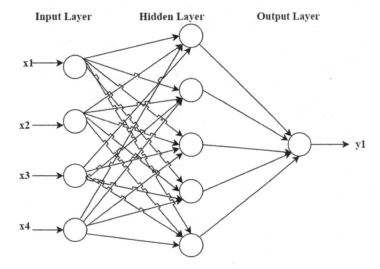

Fig. 2: Fully Connected Convolutional Neural Network.

As shown in the Fig. 2, fully connected convolution neural network consists of four inputs that is input layer, and one hidden layer with five neurons and an output neuron in output layer. In fully connected neural network each neuron of the input layer is connected all the neurons of hidden layer. Then all the computations are performed at hidden layer and the output is presented at the output layer. When any input image is fed into the convolutional layer, apply filters or choose proper parameters along with strides or padding required. Perform the operation that is convolutional on the input. For non-linearity apply Relu Activation function to the convolution operation. Pooling is performed to reduce the dimensionality. Until the error rate is reduced convolution layers are added. Then this flattened output is fed to fully connected to convolution layer. Then model will output based on the activation function used and object detection and their classification is done.

Fully connected convolutional layer is simply feed forward neural network. These layers would be the layers from last few layers connected in the network. Input to the fully connected neural network will be the output of some convolutional layer or other final pooling layer. That is output of any fully connected or convolutional neural network is a three dimensional matrix. This output will be unrolled to its values in the vector and then fed as input to another feed forward network.

Convolutional networks are simply neural networks that use convolution in place of general matrix multiplication in at least one of their layers.In convolutional network terminology, the first argument often referred to as the input(x) and the second argument was the kernel. The output is sometimes referred to as the feature map.

As shown in the Fig. 3, Convolutional neural networks are a specialized kind of neural network for processing data that has a known grid –like topology. Examples: time series data- 1-D grid taking samples at regular time intervals, image data- 2-D grid of pixels. They are tremendously successful in practical applications which indicates that the network employs a mathematical operation called convolution. Convolution is a specialized kind of linear operation.

Convolutional networks are simply neural networks that use convolution in place of general matrix multiplication in at least one of their layers.In convolutional network terminology, the first argument often referred to as the input(x) and the second argument was the kernel. The output is sometimes referred to as the feature map.

V1 is the first area of the brain that begins to perform significantly advanced processing of visual input. In this cartoon view, images are formed by light arriving in the eye and stimulating the retina, the light-sensitive tissue in the back of the eye.

The neurons in the retina perform some simple preprocessing of the image but do not substantially alter the way it is represented. The image then passes through the optic nerve and a brain region called the lateral geniculate nucleus.

A convolutional network layer is designed to capture three properties of V1:
1. V1 is arranged in a spatial map. It actually has a two-dimensional structure mirroring the structure of the image in the retina.

2. V1 contains many simple cells. A simple cell's activity can to some extent be characterized by a linear function of the image in a small, spatially localized receptive field. The detector units of a convolutional network are designed to emulate these properties of simple cells.
3. V1 also contains many complex cells. These cells respond to features that are similar to those detected by simple cells, but complex cells are invariant to small shifts in the position of the feature. This inspires the pooling units of convolutional networks.

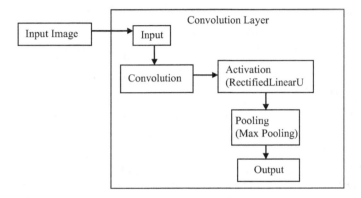

Fig. 3: Convolutional Neural Network Architecture.

Convolutional neural network exhibits Sparse Interactions that is sparse connectivity, sparse weights which are accomplished by making kernel smaller than the input image (thousands or millions of pixels, kernel – small, meaningful features such as edges – occupy tens or hundreds of pixels). Parameter sharing is nothing but the same parameters for more than one function in a model. Tied weights- because the value of the weight applied to one input is tied to the value of a weight applied elsewhere. Convolutional Neural network is used for Equi-variant representations i.e. $F(g(x)) = g(F(x))$.Convolutional neural network performs pooling. This reduces the number of parameters of the input image, by which overfitting can be reduced and extract only required features from the input. Thus reduces the computation overhead required.

To maintain or obtain the desired dimension output, padding is performed. That is kernel can move along the matrix uniformly, if the input values are padded properly with zeroes. Convolutional operation is performed to identify high level features of the image. Edges of the input image can be identified using convolutional operation. Each convolutional layer will capture low level features from the image which includes edges, pixel values that is color and so on. If the input is of 5x5x5 dimension and kernel with dimension 3x3x1, when this input is augmented with higher dimension image, then there might be some information loss, so padding is done, in which kernel can move over the matrix uniformly.

To obtain maximum values of the image portion covered by the kernel, max pooling is performed. Max pooling reduces the noise in the input along with the dimensionality reduction. Convolutional neural network will have these layers increased if the input dataset contains more complex images or to capture more low level detailed information of the image. The output of the final layer can be fed as input for other neural networks, flatten the image which is transposing the row to column vector. Then backpropagation algorithm can be applied to get the network for each training iteration. Then after several series of epochs, model will learn through experience and would provide very accurate results.

A CNN is a neural network with some convolutional layers (and some other layers). A convolutional layer has a number of filters that does convolutional operation.

In the first stage, the layer performs several convolutions in parallel to produce a set of linear activations. In the second stage, each linear activation is run through a nonlinear activation function, such as the rectified linear activation function. This stage is sometimes called the *detector stage*. In the third stage, we use a pooling function to modify the output of the layer further.

A pooling function replaces the output of the net at a certain location with a summary statistic of the nearby outputs. In all cases, pooling helps to make the representation become approximately invariant to small translations of the input. Invariance to translation means that if we translate the input by a small amount, the values of most of the pooled outputs do not change.

Invariance to local translation can be a very useful property if we care more about whether some feature is present than exactly where it is. Fewer pooling units can be used compared to detector units, by reporting summary statistics for pooling regions spaced k pixels apart rather than 1 pixel apart.

8.3.4 Image Multi Labelling

There are two types of classification in neural networks, single and multi-layer classification. Single layer classification is nothing but, only one output neuron. But with multi class classification we can have output units same as input units. Softmax activation function works by assigning the probability for the outcomes generated [3]. Then output or data is classified into class which has highest probability value. But for multi-labelling, final outcome should be independent of each other so softmax is wrong choice here. This is why sigmoid activation function is used for the final layer. The output or score values will between 0 and 1 and independent.

Another way to encode semantic relations was to use graphic representation. Using isolation logics, semantic relation can be defined over a graph which might lead into more complex graph. But this might lead into the complex of graphs. Other way is to learn the semantic relationship using labels from the data. The label relationship can be captured using either by Bayesian model and other hidden markov models can also be used. But Bayesian model is again a graph where it can be directed graph with label as node and probabilistic dependency can be represented using edges. Max likelihood function can also be used. But using these graphs it's very difficult to understand the structures.

Another method is using unified convolutional neural network and recurrent neural network for multi label classification. Using this approach semantic relationship among the labels, can be learned easily by the model. For the proposed methodology which involves more complex semantic relationship among the labels, this approach looks promising. Vectors are generated by Deep convolutional neural network for the dimensionality of the image.

8.4 Proposed Methodology

The proposed work presents a cost-effective approach for detecting the objects via images in a given environment. This approach detects presence of certain objects in the scene, regardless of their position using multi-labelling. Our proposed solution is based on Deep Learning Convolution Neural Network which makes use of LeakyRelu activation function along with BatchNormalization layer and output layer activation functions are Linear functions. Suppose if we place the camera at Vision impaired person chest, then we can calculate the distance between that particular object and the person which is through USB camera and USB laser which is integrated with the odroid system and generates an audio output. This approach is used for detecting certain objects in a given frame that is in his surroundings and calculates the distance between Visually Impaired person and the object and then provides an audio based output.

In fully connected neural network each neuron of the input layer is connected all the neurons of hidden layer. Then all the computations are performed at hidden layer and the output is presented at the output layer. When any input image is fed into the convolutional layer, apply filters or choose proper parameters along with strides or padding required. To obtain maximum values of the image portion covered by the kernel, max pooling is performed. Max pooling reduces the noise in the input along with the dimensionality reduction.

Convolutional neural network will have these layers increased if the input dataset contains more complex images or to capture more low level detailed information of the image. The output of the final layer can be fed as input for other neural networks, flatten the image which is transposing the row to column vector. Then backpropagation algorithm can be applied to get the network for each training iteration.

Then after several series of epochs, model will learn through experience and would provide very accurate results.

Perform the operation that is convolutional on the input. For non-linearity apply Relu Activation function to the convolution operation. Pooling is performed to reduce the dimensionality. Until the error rate is reduced convolution layers are added. Then this flattened output is fed to fully connect to convolution layer. Then model will output based on the activation function used and object detection and their classification is done.

As shown in the Fig. 4, Deep convolutional neural networks are nothing but feed forward networks with back propagation algorithm used to adjust the weights based on the error calculated. When compared between convolutional neural networks and deep convolutional neural networkhierarchicalpath based training in CNN which not only improves the performance also reduces the computation cost required. Weights for each convolution layer is decided during training. Convolution operation is based on Multi-layer feedback network and follows supervised techniques, but deep CNN uses both supervised and unsupervised [3].

When any input image is fed into the convolutional layer, apply filters or choose proper parameters along with strides or padding required. Perform the operation that is convolutional on the input. For non-linearity apply Relu Activation function to the convolution operation. Pooling is performed to reduce the dimensionality. Until the error rate is reduced convolution layers are added. Then this flattened output is fed to fully connected to convolution layer. Then model will output based on the activation function used and object detection and their classification is done. Activation functions like sigmoid (logistic) or soft-max is used to output as person, or table or door in our model.

As shown in the Tab. 1 below, cascade classifiers based on Haar features or Viola and Jones Haar like features are just like kernel of the convolutional neural network except that in convolutional neural network, weights or the values of the kernel will be altered based on the experience while training but in Haar the values or features are assigned or determined manually.

Output of convolutional operation is known as feature map will be converted into input vectors. Then all these vectors are combined to create a model. This vector signifies the features required. When any input image is fed into the convolutional layer, apply filters or choose proper parameters along with strides or padding required. Perform the operation that is convolutional on the input.

For non-linearity apply Relu Activation function to the convolution operation. Pooling is performed to reduce the dimensionality. Until the error rate is reduced convolution layers are added. Then this flattened output is fed to fully connect to convolution layer. Then model will output based on the activation function used and object detection and their classification is done. Activation functions like sigmoid (logistic) or soft-max is used to output as person, or table or door in our model.

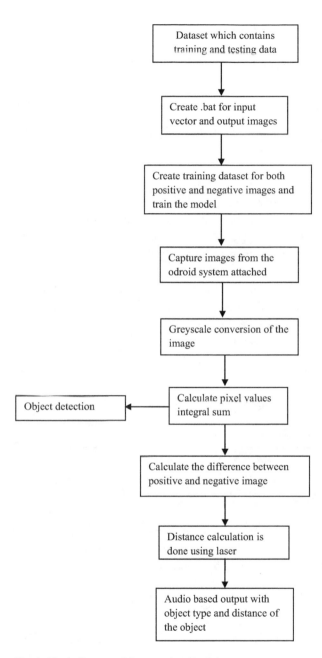

Fig. 4: Block diagram of Proposed methodology.

Activation Function used is Relu as mentioned in equation (1). This can handle non linearity of the entire network without affecting any other neurons of the convolution network layer. It performs training very fast irrespective of the input size and output layer remains the same.

$$f(x) = max(0,x) \tag{1}$$

Several others parameter of convolutional neural networks is used, such as striding, padding. Striding defines amount of shifting the kernel on the original one. Padding is like adding extra bits that is zero around the input, if the filter chosen doesn't fit input. Max pooling can be used for the purpose.

Usually Haar feature based cascade classifier is used for feature detection with limited features since all the features are manually determined. Suppose for Haar algorithms if the input is edge detection of a face and if there are some slight changes in the face then that won't be detected using haar but convolutional neural network detects it well with greater accuracy.

Haar based classifier can be used for small datasets, since there is no training required. Accuracy as shown in (2) is calculated based on the number of predictions which are correct when divided by number of examples or dataset used. The best results can be obtained when value is 1 and 0 is worst.

$$Accuracy = 1{-}ERR, \text{ where ERR is the absolute error} \tag{2}$$

Tab. 1: Comparing different Object Detection algorithms.

Parameters to be considered	Haar Cascade	CNN	Deep CNN
Accuracy	78.03%	89.02%	90.12%
Performance	0.07	0.87	0.96

Performance can be calculated using precision (equation 3) and recall (equation 4). Average value of precision is usually numerical value, which can be used to compare the performance across different algorithms. Based on precision and recall graph curve, weighted mean of each object can be calculated. Based on the ground truth, these performances can be calculated.

$$Precision = \frac{Classifier\ detecing\ correct\ deduction(TP)}{Classifier\ detecting\ correct\ deduction(TP) + Classifier\ detecting\ wrong\ deductions\ (FP)} \tag{3}$$

$$Recall = \frac{Classifier\ detecing\ correct\ deduction(TP)}{Classifier\ detecting\ correct\ deduction(TP) + Classifier\ missing\ deductions(FN)} \tag{4}$$

Time taken for computation is very high for haar based classifiers when compared to convolutional neural network and deep convolutional neural network. In convolutional neural network, the amount of pre processing required is very less when compared to haar classifier and deep convolutional neural network. Convolutional neural network is used to capture both temporal as well as spatial dependencies of the image. This architecture performs better fitting of the image because the number of parameters required is very less and no need to reuse the same weights multiple times.

In terms of complexity, haar based classifier has more complex computations since all the features needs to be explicitly determine the features. Convolutional neural network performs smaller dataset level computations since it uses labelling. Most of unsupervised learning applications use deep convolutional neural network, since it doesn't require any explicit labelling. Deep convolutional neural network can perform high complex tasks with less computation cost.

The role of convolutional neural network is to reduce the image in such a way that it can be easily processed in further steps. Input for the convolution neural network or any other neural networks which are discussed here are images, the convolutional neural network is used to perform linear processing of the image. This uses supervised learning for the purpose. If there are no labels given explicitly then based on ground truth also these computations can be performed, which is where deep convolutional neural network comes into picture. Then after these computations are performed, the output will be converted into nonlinear format that is by now the image size has been reduced and it will be in the form which can be processed easily.

Pooling can be performed to get the accuracy or the error rate will be reduced, Then the output will be very similar to input and then error rate will be calculated and update the weights. Winning neuron will be the output of the network.The work which is presented in this article is validated using various test processes [20, 21] and test techniques [22, 23].

For many tasks, pooling is essential for handling inputs of varying size. Ex: classify images of variable size, the input to the classification layer must have a fixed size. Dynamically pool features together either by using clustering algorithm on the locations of interesting features Pooling can complicate some kinds of neural network architecture that use top-down information, such as Boltzmann machines and autoencoders.

Usually, the input matrices are padded around with zero so that the kernel can move over the matrix uniformly and the resultant matrix have the desired dimension. So, P=1 means that all sides of the input matrix is padded with 1 layer of zeros as in Fig. 4 with the dotted extras around the input (blue) matrix.

The kernel moves over the matrix 1 pixel at a time. So, this is said to have a stride of 1. We can increase the stride to 2 so that the kernel moves over the matrix 2 pixels at a time. This will, in turn, affect the dimensions of the output tensor and helps reduce overfitting.

Convolution is indicated by filter (kernel). Different kernel for different dimension of input. Output obtained is the feature map(output). Padding is performed on the output to prepare it for the next layer. Just like any other Neural Network, activation function used to make the output non-linear. The function of pooling is to continuously reduce the dimensionality to reduce the number of parameters and computation in the network.This shortens the training time and controls overfitting.

After the convolution and pooling layers, our classification part consists of a few fully connected layers. However, these fully connected layers can only accept 1 Dimensional data. To convert our 3D data to 1D, we use the function flatten in Python. This essentially arranges our 3D volume into a 1D vector. The last layers of a Convolutional NN are fully connected layers.

The input is an image tensor X, with axes corresponding to image rows, image columns, and channels (red, green, blue). The goal is to output a tensor of labels $\hat{Y}$, with a probability distribution over labels for each pixel.

This tensor has axes corresponding to image rows, image columns, and the different classes. Rather than outputting $\hat{Y}$ in a single shot, the recurrent network iteratively refines its estimate $\hat{Y}$ by using a previous estimate of $\hat{Y}$ as input for creating a new estimate.

The same parameters are used for each updated estimate, and the estimate can be refined as many times as we wish. The tensor of convolution kernels U is used on each step to compute the hidden representation given the input image.

The kernel tensor V is used to produce an estimate of the labels given the hidden values. On all but the first step, the kernels W are convolved over $\hat{Y}$ to provide input to the hidden layer. On the first time step, this term is replaced by zero. Because the same parameters are used on each step, this is an example of a recurrent network.

The data used with a convolutional network usually consists of several channels each channel being the observation of a different quantity at some point in space or time. One advantage to convolutional networks is that they can also process inputs with varying spatial extents. These kinds of input simply cannot be represented by traditional, matrix multiplication-based neural networks.

Consider a collection of images, where each image has a different width and height. It is unclear how to model such inputs with a weight matrix of fixed size. Convolution is straightforward to apply; the kernel is simply applied a different number of times depending on the size of the input, and the output of the convolution operation scales accordingly.

Convolution may be viewed as matrix multiplication; the same convolution kernel induces a different size of doubly block circulant matrix for each size of input.

8.5 Conclusion and Future Work

The proposed work presents a model that is assistance device for helping visually impaired person based on image detection and classification to make visually impaired person life little easier. This model makes use of odriod system, why this system is because it can be integrated with USB cam as well laser for identifying the objects in the given environment and the distance of them from the user. So this system can be put or pinned in the chest.

The system will detect the objects within the given environment and gives the audio based output.Convolutional neural network detects the objects within the frame and computer vision provides vision to the system and works based on the training dataset used. For image sharpening or removing the outliers or getting the required features region based convolutional neural network. This helps to detect the object present in the frame with different position.

The proposed article presents the model which can be incorporated to indoor, developing the same model for outdoor environment. Instead of multi labelling, unsupervised techniques can be used with no labels using autoencoders. Limitations might be for outdoor environment, distance calculation from any moving object is quite challenging without any labels.

References

[1]   Samkit Shah, Jayraj Bandariya, Garima Jain, Mayur Ghevariya, Sarosh Dastoor, "CNN based Auto Assistance System as a Boon for Directing Visually Impaired Person", 2019 3rd International Conference on Trends in Electronics and Informatics (ICOEI), 19046658, 10.1109/ICOEI.2019.8862699

[2]   K. Visakha and S. S. Prakash, "Detection and Tracking of Human Beings in a Video Using Haar Classifier," 2018 International Conference on Inventive Research in ComputingApplications (ICIRCA), Coimbatore, India, 2018, pp. 1–4.

[3]   L. Cuimei, Q. Zhiliang, J. Nan and W. Jianhua, "Human face detection algorithm via Haar cascade classifier combined with three additional classifiers," 2017 13th IEEE InternationalConference on Electronic Measurement &Instruments(ICEMI), Yangzhou, 2017, pp. 483–487.

[4]   T. Guo, J. Dong, H. Li and Y. Gao, "Simple convolutional neural network on image classification," 2017 IEEE 2nd International Conference on Big Data Analysis (ICBDA), Beijing, 2017, pp. 721–724.

[5]   Y. Hou and H. Zhao, "Handwritten digit recognition based on depth neural network," 2017 International Conference on Intelligent Informatics and Biomedical Sciences (ICIIBMS), Okinawa, 2017, pp. 35–38.

[6]   S. Hayat, S. Kun, Z. Tengtao, Y. Yu, T. Tu and Y. Du, "A Deep Learning Framework Using Convolutional Neural Network for Multi-Class Object Recognition," 2018 IEEE 3rd International Conference on Image, Vision and Computing (ICIVC), Chongqing, 2018, pp. 194–198.

[7]   H. Jeong, K. Park and Y. Ha, "Image Preprocessing for Efficient Training of YOLO Deep Learning Networks," 2018 IEEE International Conference on Big Data and SmartComputing (BigComp), Shanghai, 2018, pp. 635–637.

[8]   W. Yang and Z. Jiachun, "Real-time face detection based on YOLO," 2018 1st IEEE International Conference on Knowledge Innovation and Invention (ICKII), Jeju, 2018, pp.221–224.

[9] M. Shah and R. Kapdi, "Object detection using deep neural networks," 2017 International Conference on Intelligent Computing and Control Systems (ICICCS), Madurai, 2017, pp. 787–790.

[10] M. A. Rashidan, Y. M. Mustafah, Z. Z. Abidin, N. A. Zainuddin and N. N. A. Aziz, "Analysis of Artificial Neural Network and Viola-Jones Algorithm Based Moving ObjectDetection," 2014 International Conference on Computer and Communication Engineering, Kuala Lumpur, 2014, pp. 251–254.

[11] J. Lin and M. Sun, "A YOLO-Based Traffic Counting System," 2018 Conference on Technologies and Applications of Artificial Intelligence (TAAI), Taichung, 2018, pp. 82–85.

[12] H. Jeong, K. Park and Y. Ha, "Image Preprocessing for Efficient Training of YOLO Deep Learning Networks," 2018 IEEE International Conference on Big Data and SmartComputing (BigComp), Shanghai, 2018, pp. 635–637.

[13] D. Peleshko and K. Soroka, "Research of usage of Haar-like features and AdaBoost algorithm in Viola-Jones method of object detection," 2013 12th International Conference on the Experience of Designing and Application of CAD Systems in Microelectronics (CADSM), PolyanaSvalyava, 2013, pp. 284–286.

[14] S. Katyal, S. Kumar, R. Sakhuja and S. Gupta, "Object Detection in Foggy Conditions by Fusion of Saliency Map and YOLO," 2018 12th International Conference on SensingTechnology (ICST), Limerick, 2018, pp. 154–159.

[15] P. P. Nair, A. James and C. Saravanan, "Malayalam handwritten character recognition using convolutional neural network," 2017 International Conference on InventiveCommunication and Computational Technologies (ICICCT), Coimbatore, 2017, pp. 278–281.

[16] D. Croce, P. Gallo, D. Garlisi, L. Giarr´e, S. Mangione, I. Tinnirello, Arianna,"A smartphone-based navigation system with human in the loop, in: Control and Automation (MED)", 2014 22nd Mediterranean Conference of, IEEE, 2014, pp. 8–13.

[17] L. Kay, "Auditory perception of objects by blind persons, using a bioacoustic high resolution air sonar", The Journal of the Acoustical Society of America *107* (6) (2000) 3266–3275.

[18] P. Bach-y Rita, S. W. Kercel, "Sensory substitution and the human–machine interface", Trends in cognitive sciences *7* (12) (2003) 541–546.

[19] K. Kaczmarek, "The Tongue Display Unit (Tdu) For Electrotactile Spatiotemporal Pattern Presentation", Scientia Iranica *18* (6) (2011) 1476–1485

[20] Naresh E. and Vijaya Kumar B.P. 2018. Innovative Approaches in Pair Programming to Enhance the Quality of Software Development. Int. J. Inf. Comm. Technol. Hum. Dev. *10*, 2 (April 2018), 42–53. DOI:https://doi.org/10.4018/IJICTHD.2018040104.

[21] Kedar Potdar, Chinmay D. Pai, Sukrut Akolkar, "A Convolutional Neural Network based Live Object Recognition System as Blind Aid", arXiv:1811.10399v1 [cs.CV], 26 Nov 2018.

[22] E. Naresh, (2020), "Design and Development of Novel Techniques for Cost Effectiveness with Assured Quality in Software Development". Ph.D thesis, Jain University. http://hdl.handle.net/10603/288589.

[23] Naresh, E., Kumar, B. P. Vijaya Kumar., Niranjanamurthy, M., & Nigam, B. (2019),"Challenges and issues in test process management". Journal of Computational and Theoretical Nanoscience, *16*(9),3744–3747.

Chethan R, Meghana D S, S Kumar

Manufacturing and Investigation of Partially Glazed Geopolymer Tile

Abstract: The present invention provides a process for the preparation of partially glazed Geopolymer tiles by using fly ash, ground granulated blast furnace slag (GGBS), alkaline solution and super plasticizers. The trial mixes are prepared by varying the mix proportion of GGBS and fly ash added with alkaline solution of specified molarity, sodium hydroxide and sodium silicate were used to prepare the alkaline solutions for the mixture. The strength of the hardened composites is tested to choose the best mix of required strength for the preparation of tiles. The super plasticizers are added to the wet mix to increase the workability of the mix. The prepared composites are placed layer by layer in the mould. The specimen is cured by air and water at ambient (room) temperature.

Keywords: geopolymer tiles, partially glazed tiles, alternative tiles, fly ash tiles

9.1 Introduction

There is a huge demand of construction material for the development of infrastructures and at the same time, the manufacturing of those shouldn't cause any adverse effect on environment. The disposal of Industries by-products are causing impact on the environment. The research work progressing on effective usage of industrial by-product to useful form under waste management system, and energy conservation.

The tile manufacturing industry is widely grown across the India. The raw material and energy consumed for the manufacturing the tiles exorbitantly increasing day by day. This project is aimed at the use of fly ash, GGBS (Industrial waste) as a major composition of tile manufacturing as an alternative to conventional cement and ceramic tiles.

The main theme of this project is to study the effective usage of Fly ash and GGBS (Industrial waste) by producing the alternative building material with less energy, resource and cost and as well as to avoid the impact on environment.

Chethan R, Department of Civil Engineering, Jyothy Institute of Technology, Tataguni, Bangalore 560082, Karnataka, India, e-mail: chethanr223@gmai.com
Meghana D S, Department of Civil Engineering, Jyothy Institute of Technology, Tataguni, Bangalore 560082, Karnataka, India
S Kumar, Department of Civil Engineering, Jyothy Institute of Technology, Tataguni, Bangalore 560082, Karnataka, India

https://doi.org/10.1515/9783110725490-009

Ordinary Portland Cement (OPC) is one of the most commonly used construction material in the world. Manufacturing of one ton OPC emits equal amount of CO_2, which is one of the major greenhouse gas. It is estimated that about 6% of total greenhouse gas is emitted by manufacturing of OPC. This result in increase in global warming, hence it has harmful effect on ecology (Davidovits. J. [1]).

In order to reduce this greenhouse gas emission, industrial by product such as fly ash, blast furnace slag, rice husk etc., can be replaced by cement as binder. Fly ash is a material which look like Portland cement both chemically and physically, which is a by-product of the combustion of coal to generate electricity at coal fired power plant.

Fly ash is abundantly available material in worldwide out of which only 17–20% was used in concrete and soil stabilization. Most of them were dumped at a waste material on a land. Fly ash mainly comprises of silicon dioxide (SiO_2) and aluminium oxide (Al_2O_3) and iron oxide (Fe_2O_3). The use of fly ash as an alternative construction material to reduce the impact on environment and to develop environmental friendly concrete by producing green concrete. Among these researches the Geopolymer was given a successful result on elimination of use of ordinary Port and cement (L. Krishnan et al., [6]).

The term "GEOPOLYMER" was first presented by Davidovits in 1978, Geopolymer is a family of mineral binders that have polymeric silicon-oxygen aluminium frame work structure which is similar to that found in zeolites. Davidovits .J in the year 1999 proposed that binders can be produced by polymeric reaction of alkaline liquids with aluminium and silicon in source materials of geological origin(Hardjito D et al., [3]).

Geopolymer products are cost effective and durable. The study of Geopolymer material properties shows very little drying shrinkage, excellent resistance to sulphate attack and good acid resistant offered when they are heat cured(J. Guru Jawahar et al., [4]). In recent research of hardened fly ash based Geopolymer have shown excellent compressive strength and elastic property, and also the behaviour of Geopolymer reinforced structural members are similar to the Ordinary Portland cement.

Geopolymer are produced by different composition, higher the concentration of solution higher will be the compressive strength (V. Bhikshma et al., [9]). The compressive strength increases with increase in fly ash content up to certain limit further increase can lead to decrease the strength in fly ash based geopolymer concrete (Kumar .S et al., [5]). Different molarity of sodium silicate solution gives different strength for different age of curing and the results gave greater value compared Portland cement (U.R. Kawade et al., [8]).

9.2 Basic Materials

Following are the list of basic materials is used for the preparation of Geopolymer concrete:

a. Class F Fly ash
b. Ground Granulated Blast furnace slag (GGBS)
c. Alkaline solution
d. Super plasticizers

Tab. 1: Physical and Chemical properties of Fly Ash.

Sl. No	Description	Values
Physical properties		
1	Specific gravity	1.94
2	Fineness (Blain's air permeability – m^2/kg)	186
3	Soundness by autoclave test (percent)	50.039
4	Residue on 45 micron sieve (percent)	39.7
Chemical properties		
5	SiO_2 (% by mass)	65.07
6	$SiO_2 + Al_2O_3 + Fe_2O_3$ (% by mass)	93.47
7	MgO (% by mass)	0.77
8	Total sulphur as sulphur trioxide SO_3 (% by mass)	0.1
9	Alkalis Na_2O	0.88
10	Total chloride	0.004
11	LOI (% by mass)	0.45

9.2.1 Class F Fly ash

Class F fly ash is obtained1from the Kadapa, thermal power plant, Andhra Pradesh state, Southern India for this experimental work. The physical and chemical properties of the fly ash are tabulated in Tab. 1.

9.2.2 Ground Granulated Blast Furnace Slag (GGBS)

Ground granulated blast furnace slag (GGBS) is a by-product of steel and iron obtained from a blast furnace in steam or water, to produce a granular, glassy product that is then dried and ground into a fine powder. The physical and chemical properties of the GGBS are tabulated in Tab. 2.

Tab. 2: Physical and Chemical properties of GGBS.

Sl. No	Description	Values
Physical properties		
1	Specific gravity	2.91
2	Fineness (Blain's air permeability – m^2/kg)	382
Chemical properties		
3	MnO (% by mass)	0.11
4	Sulphur (S) (% by mass)	0.44
5	Sulphate (SO3) (% by mass)	0.21
6	MgO (% by mass)	7.55
7	Insoluble Residue(I.R) (% by mass)	0.32
8	Chloride Content (Cl) (% by mass)	0.006
9	Glass Content (% by mass)	92
10	LOI (% by mass)	0.1
11	Moisture content (% by mass)	0.01
12	$\frac{CaO \pm MgO \pm 1/3\ Al2O3}{SiO2 + 2/3\ Al2O3}$	1.11
13	$\frac{CaO \pm MgO \pm Al2O3}{SiO2}$	1.96

9.2.3 Alkaline Solution

A combination of sodium silicate solution and sodium hydroxide were used to react with aluminium and silica in fly ash. The sodium hydroxide is in the form of flakes with 97% purity. Sodium hydroxide solution was prepared by dissolving flakes in water, for this experimental case the concentration of sodium hydroxide solution used was 6.25 molars. In order to yield this concentration one litre of the water contained 6.25 × 40 = 250grams of flakes, here the mass of NaOH solids was only a fraction of mass of NaOH solution and water is major component.

Commercially available sodium silicate was used for this experimental work with water content 33.5% and specific gravity 1.3. The ratio of sodium silicate to sodium hydroxide solution was fixed to 3.75. The alkaline solution was prepared by mixing both and sodium hydroxide solution together at least one day prior to use. The property of sodium hydroxide and sodium silicate are tabulated in below in Tab. 3 and 4.

Tab. 3: Properties of sodium hydroxide.

Appearance	Flakes
Sodium Hydroxide (% by mass)	97
Sodium Carbonate (% by mass)	2
Chlorides (Cl) (% by mass)	0.01
Sulphates (SO_4) (% by mass)	0.01
Phosphate (PO_4) (% by mass)	0.001
Iron (Fe) (% by mass)	0.005
Heavy metals (% by mass)	0.001

Tab. 4: Properties of sodium silicate.

Appearance	Clear less viscous liquid
Specific Gravity	1.35
SiO_2 (% by mass)	62
Mg_2O (% by mass)	12.1
Chlorides (Cl) (% by mass)	0.05
Sulphates (SO_4) (% by mass)	0.05
Iron (Fe) (% by mass)	0.005
Heavy metal (as Pb)	0.001
Loss on ignition (at 700 °C) (% by mass)	36
Water content (% by mass)	25.794

9.2.4 Plasticizer

Polycarboxylic ether (BASF chemicals India) is used to improve workability and to reduce water content in the Geopolymer mix.

9.3 Testing and Selection of Composites

i.   The materials such as fly ash, GGBS, sodium hydroxide and sodium silicate were taken and they were tested to determine the physical and chemical characteristics.
ii. The Geopolymer mix is prepared for different compositions to determine the strength of each composition.
iii. It is then poured into a cylindrical mould of height 30mm and diameter 30mm. It is compacted and covered by polyethene wrappers to avoid air entrapping into it.
iv. The specimen is left to dry in ambient temperature for 24 hours after which the specimens are demoulded and dimensions are measured as shown in Fig. 1, after that specimens are water cured for a period of 28 days.
v.   After 28 days of curing, the specimens are tested in Universal testing machine (UTM) shown in Fig. 2, to determine the maximum load that the specimen can withstand, and the density and compressive strength results are tabulated in the Tab. 5 and 6.
vi. The comparison of compressive strength of different composition are shown in Fig. 3.

From the test results, the blending ratio of 60:40 (GGBS: FA) was chosen as the most desirable composition. However, geopolymer tiles were also casted with 70:30 and 50:50 blending ratio.

9.4 Tile Preparation Process

9.4.1 Preparation of Alkaline Solution

i.   Sodium hydroxide flakes are weighed and mixed with water in a ratio of 1:4 and the flakes are dissolved by stirring.
ii. The solution is placed in a closed container for 30 to 45 minutes and left undisturbed until the solution gets cooled to room temperature.
iii. Sodium silicate solution is then added to the sodium hydroxide solution in the ratio of 3.75:1 and mixed well. The lid is closed after mixing and the chemicals are kept undisturbed for 24 hours.
iv. The alkaline solution should be prepared 24 hours prior to use.

Fig. 1: Determining the dimension of specimen.

Tab. 5: Density of specimens prepared for different composition.

Sl. No	Mix proportion (GGBS:FA)	Density in kg/m^3
1	80:20	2009.4
2	70:30	2004.7
3	60:40	1990.5
4	50:50	1966.9
5	40:60	1919.8
6	30:70	1858.5
7	20:80	1778.3

9.4.2 Casting of Tiles

i. GGBS and fly ash are weighed for the required proportion. The alkali solution of 50 % of total weight of fly ash and GGBS is taken.
ii. The super plasticizer is used to avoid excesses consumption of water or chemical. The quantity of super plasticizer taken is 0.8 to 0.85 % of total weight of materials.

Fig. 2: Testing of specimen.

Tab. 6: Average compressive strength of mix composition.

Sl. No	Mix proportion (GGBS:FA)	Compressive strength (N/mm²)
1	M1 – 80:20	70.73
2	M2 – 70:30	67.35
3	M3 – 60:40	77.93
4	M4 – 50:50	68.55
5	M5 – 40:60	62.50
6	M6 – 30:70	58.81
7	M7 – 20:80	53.23

iii. The lumps in GGBS and FA should be broken and the dry mix should be mixed well before adding the alkali solution.

iv. The alkali solutions and plasticizer is then added and mixed well to get a consistent paste with hand or machine.

v. When the mix attains the required consistency, mix is then transferred to the mould measuring 30 × 30 × 1 cm.

vi. The paste should be properly compacted in the mould with hands or vibrated till a levelled surface is obtained.

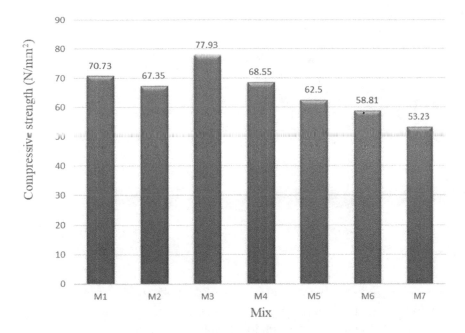

Fig. 3: Comparison of the compressive strength of the standard specimen for different compositions for 28 days of curing.

vii. The mould is then covered with a polyethene cover to avoid air entrapping and left to set for 24 hours in ambient temperature.

viii. The tile is then demoulded and water cured for 7 days. The tile is then tested to determine its engineering properties. The demoulded tile is shown in Fig. 4.

Fig. 4: Partially Glazed Geopolymer tiles.

9.5 Testing on Tile

Fig. 5: Flexural testing of tile.

1.5.1 Flexure Test

i.   The tile is tested under point loading conforms IS13630 (part 6):2006 to determine the flexural strength. As shown in above Fig. 5.
ii. The density of different composition tile is tabulated in Tab. 7.
iii. The test result such as breaking strength and modulus of rupture are for different composition of tiles are tabulated in Tab. 8.
iv. The comparison chart of breaking strength and modulus of rupture for different composition of tiles are shown in below Figs. 6 and 7.

Tab. 7: Density of tile for different composition.

Mix composition (GGBS:FA)	Density (kg/m^3)
M1 – 70:30	2047.77
M2 – 60:40	2033.33
M3 – 50:50	2028.88

Tab. 8: Breaking strength and modulus of rupture of the tile for different composition.

Mix composition (GGBS:FA)	Breaking strength (N) IS 13630(part 6): 2006	Modulus of rupture (N/mm^2) IS 13630(part 6): 2006
M1 – 70:30	1570	20.01
M2 – 60:40	1558	19.86
M3 – 50:50	1543	19.67

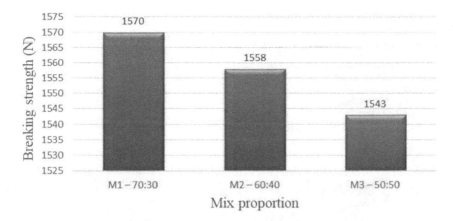

Fig. 6: Comparison chart for breaking strength of different mix proportion of tiles.

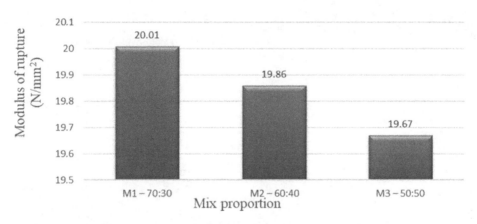

Fig. 7: Comparison of modulus of rupture for different mix proportion of tiles.

9.5.2 Moh's Hardness Test

i.   Moh's scale of hardness is determined as conforming to IS 13630 (part 13): 2006 as shown in the Fig. 8.

ii. The hardness of different composition of tile is tabulated in Tab. 9.9.

Fig. 8: Determination of Moh's scale of hardness.

Tab. 9: Moh's hardness scale of tile for different composition.

Mix composition (GGBS:FA)	Scratch hardness of surface (Moh's scale) as per IS 13630 (Part 13): 2006
M1 – 70:30	5.5–6
M2 – 60:40	6–6.5
M3 – 50:50	5.5–6

9.5.3 Water Absorption

i.   Water absorption has determined as conforming to IS 13630 (part 2):2006.
ii. The results of test is tabulated in Tab. 10.

Tab. 10: Water absorption of different composition of tile.

Mix composition (GGBS:FA)	Water absorption (%) as per IS 13630-part 2: 2006
M1 – 70:30	0.6
M2 – 60:40	0.47
M3 – 50:50	0.82

9.6 Conclusion

The fly ash and GGBS is best alternative material to produce tiles having high durability and good quality. Mix composition of fly ash and GGBS in the proportion of 40:60 respectively shows highest compressive strength of 77.9 MPa when compared to other mix. This materials strength can be corroborated with the materials used for tiles manufacturing. Tiles are prepared for three different mix proportion to compare their engineering properties such as density, modulus of rupture, hardness and water absorption. The tiles are cured by submerging in water under room temperature (26^{0} to 29^{0}C) and air dried under ambient temperature. The main concern is the energy consumptions to produce Geopolymer products as usually noticed in existing practice. Since the manufacturing process of this tiles consumes less energy, the production cost of the tiles can be reduced to 25 to 30% in comparison with vitrified and ceramic tiles. Since Moh's scale of hardness shows 6 to 6.5 this tiles can be used for industrial flooring and ware houses, where rough and smooth surfaced glazed finish tile work is expected.

References

[1] Davidovits, J. (1994). Geopolymers: man-made rock geosynthesis and the resulting development of very early high strength cement. Journal of Materials education, 16, 91–91.

[2] Davidovits, J. (1994). Global warming impact on the cement and aggre-gates industries. World resource review, 6(2),263–278.

[3] Hardjito, D., & Rangan, B. V. (2005). Development and properties of low-calcium fly ash-based geopolymer concrete. Indian standard code of practice for ceramic tiles- Methods of test, sam-pling and basis for acceptance part 2, part 6, part 13, IS 13630:2006, Bureau of Indian Standards, New Delhi.

[4] Jawahar, J. G., Lavanya, D., & Sashidhar, C. (2016). Performance of fly ash and GGBS based Geopolymer concrete in acid environment. International Journal of Research and Scientific Innovation (IJRSI), 3, 101–104.

[5]   Kumar, S., Pradeepa, J., & Ravindra, P. M. (2013). Experimental investigations on optimal strength parameters of fly ash based geopolymer concrete. Int J Civ Struct Civ Eng Res, 2, 143.

[6]   Krishnan, L., Karthikeyan, S., Nathiya, S., & Suganya, K. (2014). Geopoly-mer concrete an eco-friendly construction materi-al. Magnesium, 1, 1.

[7]   Tabassum, R. K., & Khadwal, A. (2015). A Brief Review on Geopolymer Concrete. Int. J. of Adv. Res. in Edu. Technol, 2(3).

[8]   Kawade, U. R., Salunkhe, P. A., & Kurhade, S. D. (2014). Fly Ash Based GeopolymerConcrete. International Journal of Innovative Research in Science, Engineering and Technology, 3(4), 135–138.

[9]   Bhikshma, V., KOTI, R. M., & SRINIVAS, R. T. (2012). An experimental investigation on properties of geopolymer concrete (no cement concrete).

Jyothi M. R, Dr. H. U. Raghavendra, Rahul Prasad, Priyanka B. N
Water Sensitive Urban Design

Abstract: Water Sensitive Urban Design is emerging as a crucial way of conserving water as a resource in India. Water-sensitive urban design is a land planning and engineering design approach which integrates the urban water cycle, including stormwater, groundwater and wastewater management and water supply, into urban design to minimise environmental degradation and improve aesthetic and recreational appeal. In natural environments rainwater is mostly absorbed into the ground, used by plants or evaporates back into the atmosphere WSUD aims to improve the ability of urban environments to capture, treat and re-use storm-water before it has the chance to pollute and degrade our creeks and rivers. The techniques and the methods discussed in this chapter enhance the utilization of water resources sustainably. A study in detail of Chennai Water Crisis and solutions for it, and comparing the water management in Mysore.

Keywords: pervious concrete, bio-retention, wet lands, membrane bioreactors, zero day

10.1 Introduction

Water is a valuable resource, which decides siting of an urban area. Majority of urbanization prevailed on the banks of major rivers. In 21st century where population is growing at a faster rate, people are moving into urban areas and cities for employment opportunities and good lifestyle. Supplying safe portable water for all the residents is very crucial and in today's world where value of water is at high stakes utilizing it and conserving it is very important. Undesired series of events occurred in 2019, the zero day of 2019 which led to importing of portable water from various other sources. Flooding in North Karnataka and Maharashtra are due to improper drainage system. Water sensitive Urban Design is a way of integrating hydrology cycle with the environment through proper planning. It involves all the elements of water cycle (Precipitation, Runoff, Waste water) and brings out an effective use from it (Fig. 1). This paper deals with techniques and methods used to conserve and consume water resources effectively and efficiently. The techniques discussed in

Jyothi M. R, Dept. of Civil Engineering, Ramaiah Institute of Technology, Bangalore, India,
e-mail: jyothimr@msrit.edu
Dr. H. U. Raghavendra, Rahul Prasad, Priyanka B. N, Dept. of Civil Engineering, Ramaiah Institute of Technology, Bangalore, India

https://doi.org/10.1515/9783110725490-010

this paper are Pervious Concrete, Bio-Retention, Wet lands, Membrane Bioreactors. These techniques help to – Decrease Water Pollution, control floods and provide safe portable water for residents. It reduces the Heat -Island Effect.

Fig. 1: Various elements of water cycle. (Reprinted with permission from Hack Unamatata, Smart cities: cities that have their own operating system. https://en.secnews.gr/200870/smart-cities-poleis-diko-tous-leitourgiko-sistima/) [1].

10.2 Methodology used in WSUD

10.2.1 Pervious Concrete

Pervious Concrete is a special type of concrete which is used to allow water to flow underground to replenish the underground water table. Pervious Concrete mainly consists of cement, core segregates, water, no/very-little fine aggregates and chemical admixtures (Fig. 2). The use of Pervious Concrete produces reduction in compressive and flexural strength compared to conventional concrete. The range of compressive strength for pervious concrete is between 3.5 to 28 MPa. For flexural strength, it is between 1 to 3.8 MPa. Pervious Concrete has an average seepage of 5 gallons/sq.-ft/min. It consists of 15% to 25% of more voids compared to conventional concrete. Pervious Concrete can be used for passive mitigation and active mitigation depending on the demand. There is a little amount of maintenance for proper functioning of pervious concrete. Methods used for maintenance are Pressurized Jet wash, Vacuum Sweeping. The cost of pervious concrete is 25% more than conventional concrete ranging from

Fig. 2: Pervious conceret with high infiltration capacity more than portland cement [2].

Rs. 280 per sq.-ft to Rs. 425 per sq.-ft. The major drawback pf pervious concrete is that it is effectively used where soil permeability is high.

10.2.2 Pervious Concrete

Bio-Retention is a specialized area or a pit which is used for storm water infiltration into the ground. It uses a specialized engineered soil layers and plants for effective infiltration. It can be implemented for areas not more than 5 acres. It is usually installed at residential levels as rooftop gardens, tree pits, medians etc. It helps in order to remove total suspended substances, phosphorus, nitrogen, metals, oil and grease from the runoffs collected from precipitation of a given area. Bio-Retention is created in least reduced level of a given area so that the runoff water easily gets accumulated. The engineered soil consists of fine sand and compost material at 60–70% of sand and 30–40% of compost. Bio-Retention is laid in a sequence of layers-starting from base, a gravel layer is laid, pipes are installed over the gravel layer and some more gravel is poured on to it. Then fine aggregate layer is added on top of it. On top of this, the Bio-Retention soil is added. On the surface trees, shrubs and bushes are planted in such a way that the roots grow into the Bio-Retention soil. Every time it showers, water is collected int his region and plants help to in filter the water into the ground [3]. Due to series of layers of soil added, pure water is collected through pipes and stored in collection tanks as shown below (Fig. 3).

10.2.3 Wet Lands

Wet Lands are large waterbodies within an urban area which are used mainly for storing of storm water (Fig. 4). 64% of the world's wet lands have been lost since 1900's due to overpopulation and rapid urbanization. Residential areas and commercial areas have been built upon dried wetlands, which results in flooding during heavy rains due to natural tendency of flow of water [4]. Wet Lands consist of rivers

Fig. 3: Bioretention process diagram. (Reprinted with permission from William F. Hunt., Plant selection for bioretention systems and stromwater treatment practices [9], ISBN 978–981-287-245-6. https://www.cuge.com.sg/research/CUGE-research-fellowship) [3].

and their floodplains, lakes, swamps, mangroves and salt marshes. Wet Lands help to integrate natural ecology within urban environment. It produces main source for aquatic life. Presence of Wet Lands in an urban area of proportion of 20–30% of urban area, it reduces the Heat-Island Effect as majority of water is exposed to surface. It also provides an aesthetic look for urban city. Wet Lands majorly reduce flooding during heavy rainfall by minimizing the workload on drains. It helps to replenish drinking water which can be achieved by setting treatment plants for Wet Lands. It improves urban air quality where air is often polluted due to urban

Fig. 4: Waterbodies or wetlands situated in the city. (Reprinted from https://medium.com/@coffeetrailresort/pookode-lake-18879783df22) [4].

activities. In recent years, due to improper waste management and sewage disposal Wet Lands are polluted and highly unfit for natural ecology.

10.2.4 Membrane Bioreactors

Membrane bioreactor (MBR) is a combined filtration process; it consists of both "Microfiltration" and biological treatment using activated sludge as shown in the figure below (Fig. 5). This method is best suited for both municipal works and industrial work. The above process mentioned can be used to treat domestic wastewater, it helps the wastewater color to change from blackish brown to all most colorless, and this water can be used for small urban irrigation works. Other advantages of MBRs are that it requires less maintenance compared to conventional method. It also has higher efficiency comparatively. Membrane bioreactors are used in order to install biological, chemical oxygen demands, and remove harmful pathogens in water and induce nutrients into the water which is fit for drinking and other domestic processes. The waste water collected from residential and commercial areas can be led to a nearby Membrane bioreactor plant where water is prefiltered and stored in tanks [7]. This can be used as backup at the time of water scarcity and other water crisis.

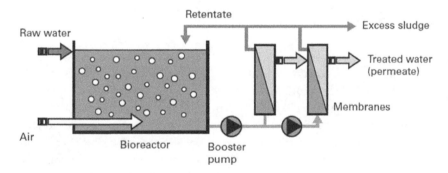

Fig. 5: Flow diagram of Membrane bioreactor (MBR) (Reprinted from https://www.suezwaterhand book.com/water-and-generalities/fundamental-biological-engineering-processes-applicable-to-water-treatment/using-clarification-membranes-in-biological-wastewater-treatment/the-main-membrane-bioreactor-families) [7].

10.3 Role of Government in Implementing WSUD

There are 4 major steps involved in implementing WSUD in a jurisdictional area is as follows. a) Framework and regulations, b) Budget and costing, c) Design and technology, d) Governance and Maintenance. There has been many advancement and research on these, but for the scope of this paper. We only consider the design aspect and maintenance of the system. Framework and Regulations relates to the role of regional and central government facilitating the implementation of WSUD in urban development and maintenance projects. It is seen that fragmentation of an area into small areas leads to good management of water, responsibilities in urban catchment management is fragmented at the same time and also integrated approaches to urban water cycle management. From past decade, improvement has led to a planned administration and maintenance work in small recognized areas. Budget and costing of is usually calculated with the concept of lifecycle assessment and outcome benefits. Using the concept of project cost evaluation, there is an assessment made regarding the efficiency and energy consumption of each sector of WSUD. Design and Technology of WSUD elements have evolved demonstrating innovation at a wide range of scales. New innovations involving collaboration with architects and engineers have extended the application of WSUD into many other dimensions related irrigation, availability of potable water etc. Governance and Maintenance are considered Very crucial WSUD to gain a wide range of political support and public support. It is elementary to provide enhanced implementation, as well as improving the technical capacity in urban environments. It also concerns with the problem faced and community acceptance of the WSUD new projects.

10.4 Case Study

Chennai faced water deficit on 19th June, 2019. It was quoted as Zero Day where almost no water was left. This is an effect due to the fall of monsoon since past two years shown in Fig. 6. 2017 and 2018 was a bad year for southwest monsoon. Rainfall drop since 2017 from 140 cm annually to 83 cm in 2018. Due to failure of monsoon in 2018 and improper water collection reservoirs, Chennai faced the water crisis. Chennai goes through a sequence where it gets flooded once in every 5 years. This is due to heavy southwest monsoon. Water has to be effectively stored in form of underground water or in storage tanks when there is excess precipitation. As Chennai is situated on the coast of Bay of Bengal majority of storm water is let into the sea, which could have been stored after filtration or replenished by underground water. Chennai has a total area of 426 sq.km with a population of 70.9 Lakhs. Considering 135 liters per capita per day as average water consumed, it would require 2,83,50,000 cu.m of water per month to satisfy Chennai's needs [8]. This would require 130 mm of rainfall per month. Chennai only receives peak rainfall from the month of October to December. The major rivers of Chennai are Cooum, Adyar and Kortalaiyar. These have been polluted since past decade. Due to industrial activities and draining of waste water, the water from these rivers is unfit for drinking and sanitation purposes. Chennai majorly depends on four reservoirs and lakes namely, Madhurathakam,

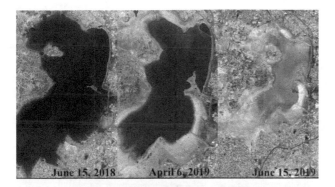

Fig. 6: Satellite image of Chennai state during 2018 June, 2019 April and June. (Reprinted from https://www.asianage.com/photo/in-other-galleries/210619/drying-up-before-and-after-satellite-photos-of-chennais-lakes.html) [8].

Pulicat, Kaliveli and Sholavaram for portable water. During the drought all the four lakes were dried up and there were no signs of underground water.

In order to avoid any future water crisis in Chennai, it is very important to replenish underground water as a storage unit. This is because fluctuation of the monsoon every year, water stored in reservoirs may run out easily. So, underground water plays a key role. In the western part of Chennai, on the banks of river Cooum, there is an abundance of deep loamy soil which has infiltration rate of 10–20 mm per hour. Bio-Retention can be setup here in order to increase the underground water table effectively (Fig. 7). Chennai does not have very good soil for infiltration other than this area, majorly covered with clay soil and calcareous clay soil which is unfit for infiltration.

The area selected for Bio-Retention is at the bank of river which is suitable for collection of storm water as it is at a lower reduced level compared to the neighbouring region.

In order to carry this out, there are three methods to be followed. The methods are:
- Rain water Harvesting, harvesting in individuals' households, and commercial rooftops and parking lots can be used for cleaning purposes and flushing, carwashes etc.
- Bio-Retention method, it is very important to relay on the ground water for Chennai
- Membrane Bioreactor

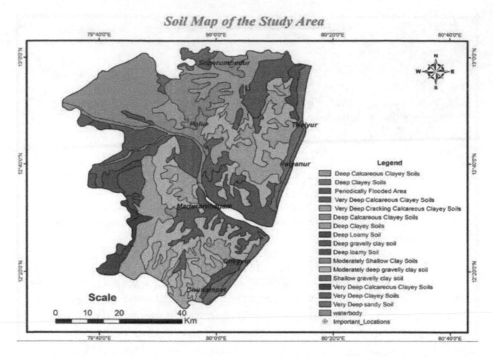

Fig. 7: Remote sensing image of Chennai state of soil conditions. (Reprinted with permission from Muthuswamy seeniranjan, Study and analysis of chennai flood 2015 using GIS and multicriteria technique. Journal of geographic information system [6] (9):126–140. https://www.researchgate.net/figure/Soil-map-of-the-study-area_fig2_317012219).

10.5 Comprasion Study (Rainfall Analysis Mysore)

Mysore district receives an average rainfall of 776.7 mm. Mysore consists of 53 rainy days in the district. The Majority of Rainfall is caused by southwest monsoon (Fig. 8 a). The rainfall reduces over time from West to East.The average temperature of 27°C annually. Water conservation and Artificial Recharge: Large parts of Mysore and T. Narasipura taluks, Nanjangud and H.D. Kote.The west part of the district, where the topography is hilly and elevated, artificial recharge structures like gully plugs contour bunds and contour trenches are constructed. But in plain areas the water is percolated into the ground, as the soil is good for percolation. Due to the above-mentioned factors there is no need for new WSUD system to be implemented at present. May be due to decline in monsoon rainfall (Fig. 8 b) and over exploitation, Mysore may need WSUD new innovations in future.

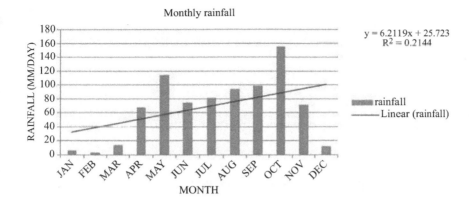

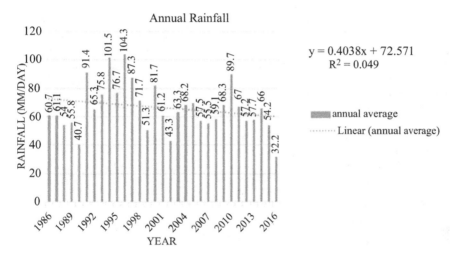

Fig. 8 a, b: Graph shows monthly and annual trend of rainfall pattern for Mysore over years. (Reprinted with permission from Malini, Climate change and climate variabilty for Mysore taluk using RS and GIS. IRJET Volume:05 Issue: 02/02/2018 [5] https://www.irjet.net/archives/V5/i2/IRJET-V5I2256.pdf).

10.6 Conclusion

The concept of water sensitive urban design is based on developing plans to have a inbuilt system to integrate the water cycle, with proper utilization and proper sustainable design. In order to utilize the storm water properly in urban landscape and topography, there is need of a system like WSUD. The use of WSUD, espouses the need to involve storm water management by planning and designing of urban areas and then apply it to a larger scale that is for the whole catchment area for maximum utilization of stormwater. Opportunity to improve the integration of stormwater management, and functions within the urban landscape designs at a large scale within the public realm and private buildings were presented in this paper. The next step is to adopt the concept of on-site water management and stormwater management which reduces the cost for new setup.

References

[1]   Andrew Speers and Grace Mitchell, Integrated Urban Water Cycle, Commonwealth Scientific and Industrial Research Organisation (CSIRO), Australia.
[2]   Ground Water Information Booklet Mysore District, Karnataka Government Of India Ministry Of Water Resources.
[3]   Lanscape-Institute, http://www.youtube.com/watch?v=b_DTnOzYTR4[/embedyt], unpublished
[4]   Madras.com, "https://www.madras.com/v/geography/" for topo sheets and soil types.
[5]   Malini, Ravikumar, Dr. Gouda, "Climate Change and Climate Variability for Mysore Taluk using RS and GIS", International Research Journal of engineering and Technology.
[6]   Muthuswamy Seeniranjan, (2017). Study and analysis of chennai flood 2015 using GIS and multicriteria technique. Journal of geographic information system, 9,126–140.
[7]   R. R. Brown, N. Keath, Urban water management in cities: historical, current and future regimes, Victoria, Australia.
[8]   T. H. F. Wong, An Overview of Water Sensitive Urban Design Practices in Australia, Ecological Engineering, PO Box 453, Prahran, Victoria 3181, Australia.
[9]   William F. Hunt, Bill Lord, Benjamin Loh, (2015). Plant selection for bioretention systems and stormwater treatment practices, 978–981-287-245-6, Centre for Urban Greenery & Ecology, Singapore.

Jyothi. M. R, Dr. H U Raghavendra, Dr. Siddegowda

Degradation of Groundwater Resources By Waste Exposed Land Use In Urban Cities

Abstract: Urbanization has led to the generation of large quantity of domestic solid waste. A part of it is properly managed and the rest is dumped illegally around societal community varies the natural resources of groundwater and the soil which has hazardous impact on biotic environment. Analysis of the samples are carried out for assessing the impact of solid waste on groundwater characteristics and check there permissible limits are within the following guidelines of WHO and BIS, if not proper segregation of solid waste need to be done at the source, minimizing the generation of waste and improving the methods of handling it. In this chapter discussed the degradation of Groundwater Resources By Waste Exposed Land use In Urban Cities

Keywords: groundwater, solid waste, heavy metals, segregation, biotic environment, environmental impact

11.1 Introduction

During the early interlude, solid wastes were unremarkably disposed off, as the density of population was low with large open space. It is been established that in 2016, 2.01 billion tonnes of waste has been generated which is equals to 0.74 kilograms/person/day. This may increase at a higher level of waste of 3.4 billion tonnes i.e., 70% of increase from 2016 to 2050 through drastic urbanization.

Important cities such as Mumbai and Delhi of India generates a huge amount of SW among other metro cities. CPCB reports have evaluated that the greater Mumbai and Delhi produces per day 11,000 and 8,700 tonnes of solid waste.

There are diverse groupings of solid waste produced, each type of litter take their time to degeneration as shown in Tab. 1.

However this familiar practice allows decreasing the waste inflow amount to the natural water bodies and open dumps [10] it is also an occupational risks and hazardous [12]. Therefore, concerning waste open dumping and open burning, explores environmental impacts due to unmanageable solid waste causing environmental pollutions. The Fig. 1 below shows a schematic representation of a review report.

Jyothi. M. R, Dr. H U Raghavendra, Dept. of Civil Engineering, Ramaiah Institute of Technology, Bangalore, India, e-mail: jyothimr@msrit.edu
Dr. Siddegowda, Dept. of Civil Engineering, SJC Institute of Technology, Chickaballapur, India

https://doi.org/10.1515/9783110725490-011

Tab. 1: Different categories of waste generated, *edugreen.teri.resi.in/explore/solwaste/types. htm.

Type of litter*	Approximate time it takes to degenerate the litter*
Organic waste such as vegetable and fruit	– week or two
Paper	– 10–30 days
Cotton cloth	– 2–5 months
Wood	– 10–15 years
Woollen items	– 1 year
Tin, aluminium and other metals items such as cans	– 100–500 years
Plastic bags	– one million years?
Glass bottles	– undetermined

One of the widely spread natural resource is ground water. In order to make use of the available ground water source in an optimal way, the precious resources has to be well understood in terms of its occurrence, behaviour and quality. Quality of water is a function of its physical, chemical and bacteriological characteristics which depends upon many factors like natural and anthropogenic activities. Unmanaged and uncontrolled solid wastes dumped openly on land produces liquid waste and gaseous emissions creating environment pollution and also causing breeding for diseases- bearing animals and microorganisms as shown in Fig. 2. The production of solid waste from various sources such as households, markets, industries shops and abattoir results in improving the living standards of the people. These wastes can also contaminate groundwater [5]. Without careful judgement of land and dumping the waste and leachate on landfills contaminates surface and ground water supplies and land resources [2]. Shallow wells are more dangerously polluted at certain period when large amount of solids, microbial organisms are present in leachates [5]. Many measures are showed through research paper on waste disposal contaminant on environment. The present study is aimed on measuring the impact on the quality of ground water from solid waste dumping on surrounding area.

11.2 Case Study

11.2.1 Mumbai

As per the rules the landfills sites must be kilometres away from the nearest human civilization but they are so very near to the colonies in Mumbai.

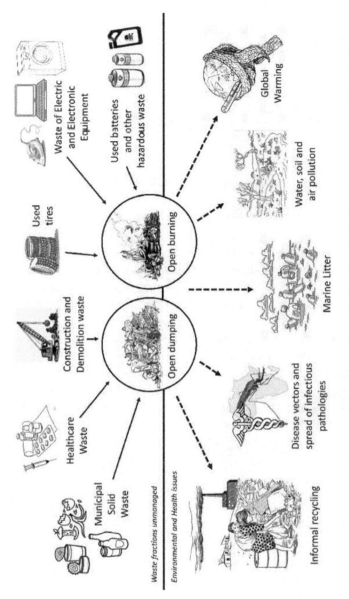

Fig. 1: Source of contamination due to SW negligence (Int. J. Environ. Res. Public Health 2019, 16, 1060; doi:10.3390/ijerph16061060 www.mdpi.com/journal/ijerph.).

Fig. 2: Waste exposed in surrounding areas of urban cities.

In Mumbai district since last four decades CGWB is monitoring the wells. The main objective is to check the quality of groundwater in Mumbai district. The samples were collected and immediately analysed for the quality parameters of groundwater in Regional Chemical Laboratory of the Board at Nagpur. The parameters analysed includes Fluoride, Nitrate, pH, EC, Total hardness and Total Alkalinity. The ground and surface water pollution is one of the major pollution due to dumping land of sewage and Industrial effluents which creeks into water bodies. The concentrations of Hg is 1.90 mg/l as found out from MPCB (Maharashtra Pollution Control Board). The Dye industries and Alkalis most responsible for the discharge of pollutants in Thane streams. In Thane and Chembur, the concentration of Arsenic was found high in fishes was observed 2 mg/l. The other Heavy metals like Cadmium found to be 12.60 mg/l, Lead-0.60 mg/l, Cu- 8.84 mg/l was noticed in streams of Maharashtra, 2013 CGWB.

11.2.2 Delhi

Extensive use of electronical gadgets like mobiles, tablets, laptops, smart watches has made E- waste cause problems on environment as shown in Fig. 3. based on a new study, unused devices results in enormous storage of Electronic waste, which is decreasing the quality of soil and ground water in the national capital.

Ground water samples were analysed for heavy metal contamination, such as Pb, Cd, Cu and were found 20 times higher than the standards by CPCB (Central Pollution Control Boards) from the area. Nickel and Chromium were about five times higher than that of the standards [9].

Many places in Delhi contamination of groundwater is due to numerous factors, like poorly constructed UGD system, construction activities (cement, paints, varnishes etc) and leakages in landfills (Fig. 4).

Fig. 3: Dumpingofewasteonsoilandgroundwater(http://thehindu.com/news/cities/Delhi/arti
cle236710.ece).

Fig. 4: Contamination of groundwater due to dumping of various waste materials in Delhi
(https://rainwaterharv.weebly.com/factors-that-effect-water-quality1.html).

11.2.3 Chennai

In Chennai, India, more than 3200 tonnes per day of solid waste are generated,
heavy metals leaches in to water causing health hazard to people. At various depths
of 2.5 to 5.5 m of soil samples the concentration of heavy metals ranges from
3.78 mg per kg to 0.59 mg per kg, sandy clay layer showing high concentration in
the top soil to 5.5m depth. Representing the influence of the dumping activities, the
heavy metal concentration showed less with increasing the soil depth.

11.3 Study Area

11.3.1 Bangalore

Drastic Horizontal growth of Bangalore exceeded the milestone of 13°05′06″N – 78°0′0″ E from the establishment of International Airport, IT sectors and apartment dominances under Hebbal constitution. However, with a fast growing population and tremendous necessities of the IT sector, the local authorities are not able to provide the necessary services like water supply, road maintenance and solid waste management etc., to full satisfaction [13]. Hebbal area is located in the North zone of Bangalore. Hebbal lake is

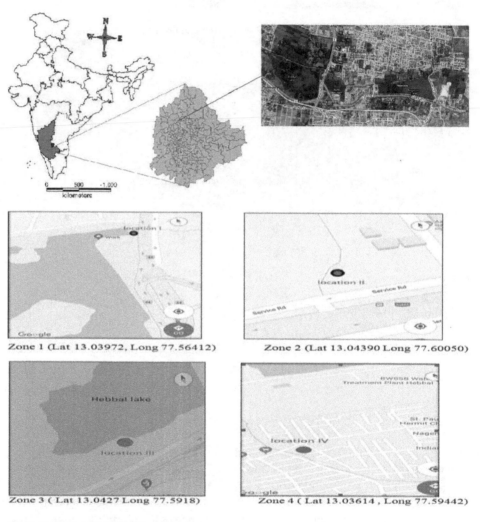

Fig. 5: Location selected for the study area.

one of the third oldest lakes in Bangalore. The catchment area of the lake was found to be 37.5 km2 (Fig. 5). In 1974 the lake area was 7.795 km2 and in 1998 it was 5.775 km2.

Physico- chemical and biological properties of the lake has altered the acceptable standards due to sewage inflow and solid waste dumping in to the lake.

11.4 Research Methodology

The derivation of any database is usually based on periodic and routine monitoring. The MSW composition and generation are totally based on the sampling method [1]. Even the responsible authorities use the time series to predict or to estimate the MSW waste generation and its composition [6]. To find the changes occurring insight the groundwater and reach the specified objectives, groundwater samples were analysed for the quality purpose (Fig. 6).

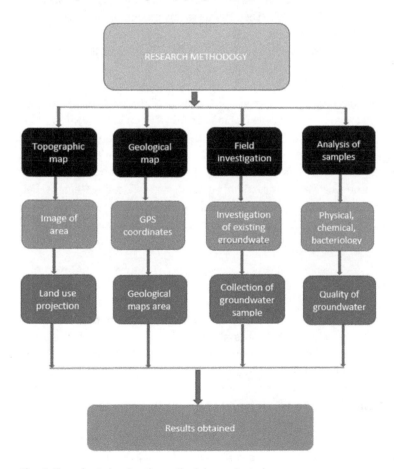

Fig. 6: Flow chart showing the methodology adopted.

11.5 Experimental Results

Tab. 2: Analysis of ground water samples.

Sl. No	Stations	Odour	pH	Acidity	Alkalinity	Total hardness	Ca++	Mg++	TS	DO	BOD	COD
1	Zone1	–	6.7	55	348	320	218	102	671	6.1	27	267
2	Zone2	–	7.2	65	335	358	243	115	629	5.4	25	460
3	Zone3	–	7	67	375	659	514	144	679	7.4	26	143
4	Zone4	–	6.8	46	314	347	240	115	706	7.4	26	181

The analysis of various physico- chemical and bacteriological parameters of groundwater sample around the dumping area are provided (Fig. 5). Various examinations such as odour, pH, Alkalinity, Acidity, Total solids, Total hardness, Magnesium hardness, Calcium hardness, DO, BOD, COD are analysed and results are tabulated in Tab. 2. All the samples for odour test done were clear and unobjectionable. The pH values for all the samples at different locations are within the permissible limit. The range of acidity in groundwater samples are also tabulated as shown in (Tab. 2), it is been observed that acidity is less in Zone4 and was more in Zone1, Zone2, Zone3 regions. The range of alkalinity, total hardness, total solids were also found to be within permissible limits of WHO and BIS. Were as Ca++ hardness, Mg++ hardness were found more than the permissible limits and DO was also found more than the limit for different zones. The BOD was found within the permissible limits in all the zones whereas COD showed an increase level in Zone1 and Zone2. This might be one of the reason where solid waste dumped in the area is releasing leachate that causes pollution. A graph (Fig. 7) has been plotted to show the variations of each parameters at every zones selected from the study area for analysing the groundwater samples.

11.6 Conclusion

The Contamination of groundwater is the major risk on environment due to un-engineered land filling of solid waste. The analytical results of the hydro-chemical composition of the groundwater in the study area showed that acidity, total hardness, dissolved oxygen chemical oxygen demand to be slightly high. occurrence of higher concentration of all the above parameters suggesting that the groundwater sample collected from dump site is not suitable for human consumption. Crude dumping of solid waste in open spaces which constantly block both primary and secondary drainage networks and pools of stagnant water. Solid waste generates a liquid called leachate which not only contaminate surface water but also the groundwater.

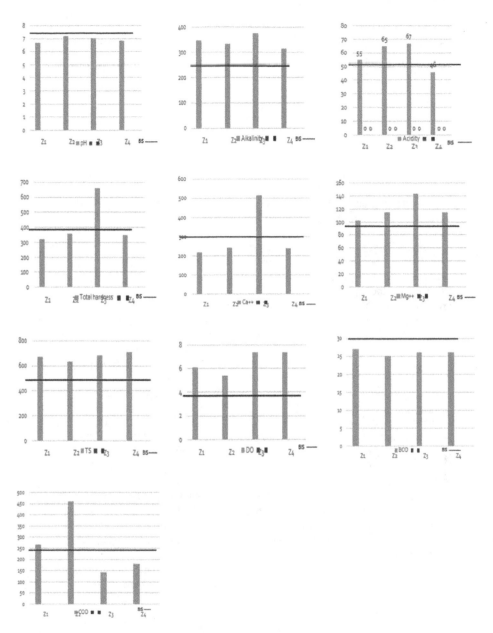

Fig. 7: Showing variations of all the ground water quality parameters at the study area.

To overcome the pollution of ground water from seepage of toxic liquid of solid waste dumping, leachate treatment plant need to be installed, monitored, treated and disposed off in proper manner to the environment, hence decentralized segregation at the source need to be adopted.

The decentralized approach could be one effective methods to solve the problems of solid waste management in Hebbal as it has potential to reduce the contamination of ground water and quantity of waste through the seepage of leachates and amount of air pollution.

References

[1]   Beede, D.N. and Bloom, D.E. (1995). The Economics of Municipal Solid Waste. World Bank Research Observer, 10, 113–150.

[2]   G I Parvathamma, (2014). An Analytical Study on Problems and Policies of Solid Waste Management in India –Special Reference to Bangalore City, IOSR Journal of Environmental Science, Toxicology and Food Technology, 8, 06–15.

[3]   Gutberlet, J.; Baeder, A.M.; Pontuschka, N.N.; Felipone, S.M.N.; dos Santos, T.L.F. (2013). Participatory research revealing the work and occupational health hazards of cooperative recyclers in Brazil. Int. J. Environ. Res. Public Health, 10, 4607–4627.

[4]   Linzner, R.; Salhofer, S. (2014). Municipal solid waste recycling and the significance of informal sector in urban China. Waste Manag. 32, 896–907.

[5]   Meadows, R. (1995) Livestock Legacy Environmental Health Perspectives 103 {12} 1096;1100.

[6]   Mritunjay Kumar,(2017). Decentralization of Solid waste management in Delhi: an opportunity for revenue generation, Pacific Business Review Int., 2, Aug.

[7]   Parameswari, K.; Padmini, T.K.; Mudgal, B.V. (2015). Assessment of soil contamination around municipal solid waste dumpsite. Indian J. Sci. Technol., 8, 36.

[8]   Pavan, H B Balakrishna, (2014). Decentralized composting of municipal solid waste in Bengaluru city. Int. Journ. Of Research in Engg. And Technology,03, 6, June.

[9]   Rashmi Makkar Panwar and Sirajuddin Ahmed (2018). Assessment of contamination of soil and ground water due to e-waste handling. Current Science, 114 (1), Jan, 166–173.

[10] Sasaki, S.; Araki, T. (2014). Estimating the possible range of recycling rates achieved by dump waste pickers: The case of Bantar Gebang in Indonesia. Waste Manag. Res. 32, 474–481.

[11] Shekhar, S., Purohit, R.R. and Kaushik, Y.B. (2012). Groundwater Management in NCT Delhi. Central Ground Water Board, Govt of India, New Delhi.

[12] Singh, M.; Thind, P.S.; John, S. (2018). Health risk assessment of the workers exposed to the heavy metals in e-waste recycling sites of Chandigarh and Ludhiana, Punjab, India. Chemosphere, 203, 426–433.

[13] T V Ramachandra and Shruthi Bachamanda, (2007). Environmental audit of Municipal solid waste management, Int. J. Environmental Technology and Management, 7, 3/4.

Santhosh D, Nambiyanna B, Harish M L, DR R Prabhakara

Use of Effective Placing Different Lateral Load Resisting Structural System in High Rise RC Frame for All Seismic Zones By Using Response Spectrum Method

Abstract: Developments of industries are fast in developing countries with fast growing construction industries.. In present situation population density is goes on increasing which is affecting the real estate that is the demand is increasing rapidly, to overcome these problems tall building construction is one solution. As the nature of the structure expands the effect of wind and earthquake also increases due to this the buildings may failure or building will displace in larger amount, to make building safer against earthquake load different types of lateral applied load resisting systems are used for construction.

In the present study we have considered a structure of size (30X30m) with 6 bays in each direction and different types of lateral load resisting structural systems like Bare frame model, core wall structural system, shear wall structural system, outrigger structural system, outrigger with belt truss structural system, conventional brick infill wall system and Aerated light weight concrete block infill structural system. All these systems were modelled using ETABS 9.7.4 in all seismic zones with different heights ie 120meters (40storeys), 180 meters (60storeys) and 270 meters (90storeys). Response spectrum method is carried out with considering earthquake forces andobtained results are tabulated with respect to storey displacement, storey drift and base shear. After comparing the all types of structural forms efficient one is identified.

12.1 Introduction

Tall structures are the structures whose height will make more sense as compared to other parameter. As the structure height increases the effect of Earthquake force also increases, the lateral load makes structures unstable and people feel uncomfortable

Santhosh D, Department of Civil Engineering, Ramaiah Institute of Technology, Bengaluru, Karnataka, India, e-mail: santhu4098@gmail.com
Nambiyanna B, Harish M L, Department of Civil Engineering, Ramaiah Institute of Technology, Bengaluru, Karnataka, India
DR R Prabhakara, Brindavan Collage of Engineering, Bangalore

https://doi.org/10.1515/9783110725490-012

to live. Structures are behaves like a cantilever beam, as the height increases the overturning movement also increases, than structure starts bending or it deflects for certain amount, tall structure should be safe against shear and bending.

When the building resisting shear loads it should not fail by shearing off and the deflection or later displacement must be within elastic regain.When the structure is affected by wind and earthquake loads it creates an overturning moment on the structure due to this moment the building may overturn or columns are going to fail by compression or tension or it deflect for large amount. Tall structure should be safe against bending effects,Tall structures must be safe against shear and bending effects, to overcome the lateral effects different types of lateral load resisting systems are used for high rise construction they make structure stable and make people feel comfortable. Different types of lateral load resisting systems used are Braced frame, Core wall system, Outrigger and belt truss system, shear wall, framed tube, Core braced and Infill wall structural system.

12.1.1 Shear Wall system

The shear walls are generally placed from bottom to top height of building.it may be brick or RC wall. In high rise buildings the thickness of the shear walls can be 150mm or 400mm depends on requirement. The shear wall beams have vertical orientation that carry earthquake load in downward direction.

12.1.2 Outrigger Structural System

Framework is a compelling horizontal burden opposing framework where in outer segments are associated with focal center divider with profoundly firm outriggers at deifferent levels. The solidness in external segments elevates the obstruction against upsetting minutes. This auxiliary structure involves a focal center with supported edge/shear divider with cantilever bracket/brace known as Outrigger bracket. With center halfway found outrigger stretches out on the two sides or it might be on one side of the structure, from the literature review we observed that the outrigger system is placed at the 3/4th height of the structure is most efficient. Present study we also placed outrigger system at 3/4th height of the structure.

12.1.3 Infill Wall System

Reinforced concrete (RC) frame working with workmanship infill dividers have been broadly built for developed for business, mechanical and multi-story private uses in seismic locales. Stone work infill ordinarily comprises of blocks or solid squares built among pillars and sections of a strengthened solid edge. The brick work infill boards are commonly not considered in the structure procedure except their self weight in analysis what's more, treated as building (non-auxiliary) parts. Appropriately planned infills can expand the general quality, horizontal obstruction and vitality dissemination of the structure. An infill divider lessens the sidelong avoidances and bowing minutes in the casing, in this way diminishing the likelihood of breakdown. Consequently, representing the infills in the examination and configuration prompts thin edge individuals, diminishing the general expense of the basic framework.

The structure with shear wall has more base shear due to increase in mass, due to high base shear it has less time period and has less lateral drift as compared other systems(Muralidhar G B, Swathi rani K S [4]).

For 20 storied building with belt truss system at 0.46 times of the overall height of the building produces less drift and for 25 storied building at 0.5 times the height of the building (Akshay A. khanorkar, S V Dange.[1]).

For dynamic loadings, shear wall core reduced the lateral drift by 29% as compared bare frame; shear wall with each side of external bay reduces the floor drift. Shear wall at center and on each side of external bay are preferable.

ALC infill as shear wall has a less lateral displacement and storey drift as compared to bare frame, ALC block infill has a significant effect on the performance of structure(Vidhya P. Namboothiri [11]).

12.2 Objectives of Present Work

- Three types of lateral load resisting systems for a tall structure are analysed against earthquake loads and all parameters mainly storey drift,lateral displacement and base shear are compared for all earth quake zones.
- The performance of optimized location of outrigger with belt system and without belt system are analysed for all earth quake zones.
- Two types of masonry infill systems, conventional brick and Aerated light weight concrete block are considered and analysed for all earth quake zones.
- 3D models of systems considered for present study are Bare frame, core wall, shear wall system, masonry infill system, outrigger with belt truss and outrigger without belt truss system, are analysed for all earthquake zones and the results are compared.

12.3 Methodology

In the present study, the 3D models of seven types of systems with 40,60 and 90 storeys are modelled in all seismic zones to study and compare the structural performance.All models have same section and material properties as mentioned below. All models are analysed in X-direction by response spectrum method

Tab. 1: Sections and material properties considered for present study.

Section Properties	
Number of floors	40/60/90
Number of bays in X and Y direction	6
Size of bay	5m
Height of each floor	3m
Beam size	300x600mm
Column size	1200x1200mm
Slab thickness	175mm
Infill thickness	230mm
Size of outrigger and belt truss system	300x600mm
Thickness of shear wall	300mm
Concrete grade	M40
Steel grade	Fe500
Concretedensity	$25kn/m^3$

Tab. 2: Properties of infill materials.

Properties	Conventional brick	ALC Block
Density (kn/m^3)	19.2	6.5
Compressive strength (mpa)	4.3	3
Youngs modulus (mpa)	2640	1840
Poisons ratio	0.16	0.25

12.4 Response Spectrum Analysis

Reaction spectra are exceptionally helpful instruments of seismic tremor designing for examining the exhibition of structures and hardware in quakes, since numerous essentially as basic oscillators. Subsequently in the event that you can discover the characteristic recurrence of the structure, at that point the pinnacle reaction of the structure can be evaluated by perusing the incentive starting from the earliest stage range for the proper recurrence. In most construction regulations in seismic locales, this worth structures the reason for computing the powers that a structure must be intended to oppose seismic examination. For different zones on the basis of zone factors(i.e.0.10, 0.16, 0.24 and 0.36) for zone2, zone3, zone4 and zone5 respectively.

The design base shear (Vb) compared with response spectrum base shear (Vba) calculated by using a time period Ta.

12.5 Models Considered for Present Study

Seven types of lateral load resisting systems are considered for analysis and are modeled for 40, 60 and 90 stories. Bare frame system, core wall, peripheral shear wall, outrigger, outrigger with belt trus, conventional brick infill and ALC block infill system. All models are modeled using Etabs 9.7.4 software andanalysed by response spectrum method for all seismic zones. Bare Frame represent the below fig.1.

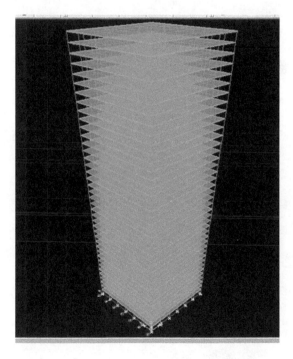

Fig. 1: Bare frame.

12.6 Results and Discussion

In this chapter the results of the analysis conducted for 3D RC frame systems with different height by varying number of storeys from 40, 60 and 90 storeys. Response spectrum analysis is carried out for all these different structural systems considering all seismic zone and Medium soil as per Indian Standard code IS-1893(Part1:2002). The

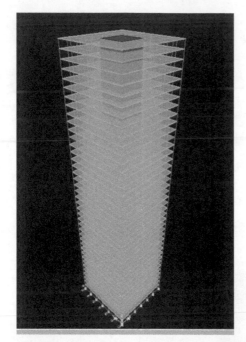

Fig. 2: Core wall.

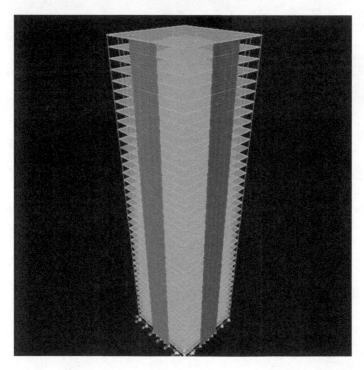

Fig. 3: SHear wall.

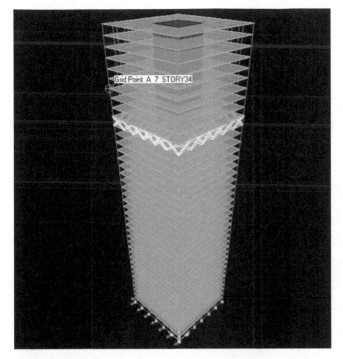

Fig. 4: Outrigger.

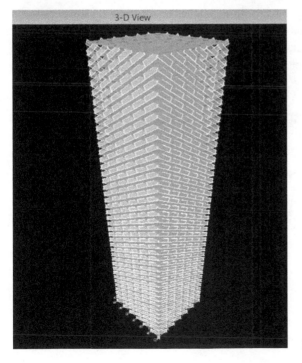

Fig. 5: Outrigger with belt truss.

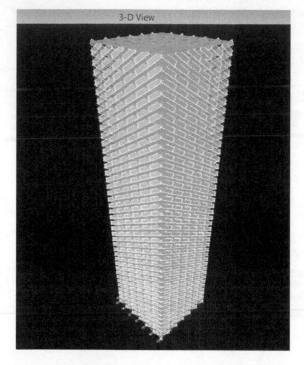

Fig. 6: RC frame with Infill wall.

parameters considered for comparative study were Maximum lateral Displacement, Storey Drift & Base Shear. The following figures showing the comparison of 40,60 and 90 storey structures with all seismic zones are drawn and results are compared.

12.6.1 Lateral Displacement

Following figures showing the lateral displacement of 40,60 and 90 storey for all structural systems in all seismic zones

From the above figures it is observed that lateral displacement is maximum for Bare-frame structural system and it is maximum in seismic zone 5, and the lateral displacement is minimum for outrigger with belt truss system for all number of stories and seismic zones.

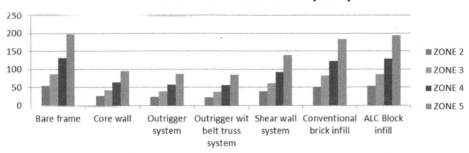

Fig. 7: Showing the lateral displacement of 40 storey models for all seismic zones.

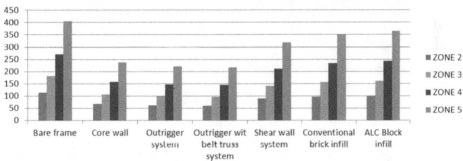

Fig. 8: Showing the lateral displacement of 60 storey models for all seismic zones.

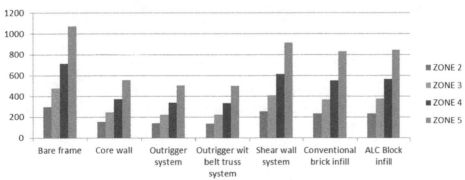

Fig. 9: Showing the lateral displacement of 90 storey models for all seismic zones.

12.6.2 Storey Drift

Following figures showing the lateral displacement of 40,60 and 90 storey for all structural systems in all seismic zones. But the storey drift values given by the software are in storey drift ratio i.e. ratio of storey drift and floor height.

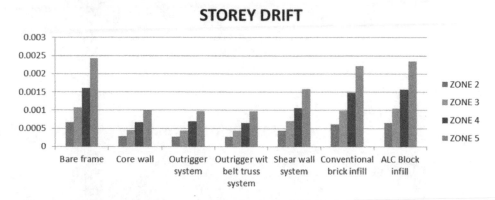

Fig. 10: Showing the storey drift of all models of 40 storey in all seismic zones.

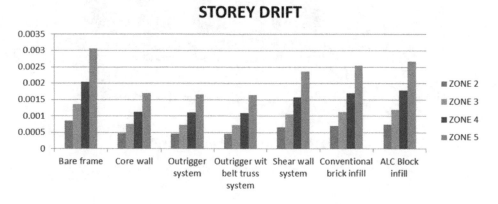

Fig. 11: Showing the storey drift of all models of 60 storey in all seismic zones.

From the above figures it is observed that storey drift is maximum for Bareframe structural system and it is maximum in seismic zone 5, and the storey drift is minimum for outrigger with belt truss system for all number of stories and seismic zones.

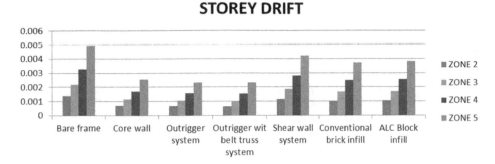

Fig. 12: Showing the storey drift of all models of 90 storey in all seismic zones.

12.6.3 Base Shear

Following figures showing the lateral displacement of 40,60 and 90 storey for all structural systems in all seismic zones

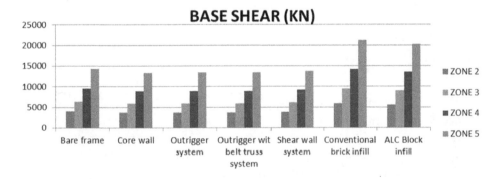

Fig. 13: Showing the Base shear of all models of 40 storey in all seismic zones.

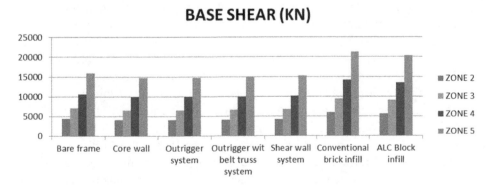

Fig. 14: Showing the Base shear of all models of 60 storey in all seismic zones.

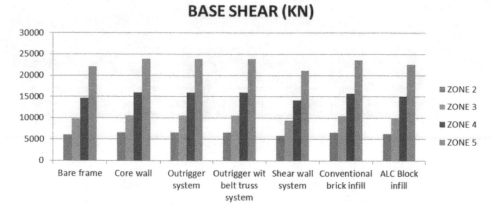

Fig. 15: Showing the Base shear of all models of 90 storey in all seismic zones.

From the above figures it is observed that base shear is maximum for conventional brick infill in seismic zone5 due to increase in self weight of the structure, and it is minimum for core wall structural system for all number of stories and seismic zones. Refer above fig. 15.

12.7 Conclusion

1. The maximum storey displacement for a 40 storey RC frame structure varies in accordance with the different structural forms adopted. Lateral displacement was reduced by 52% with core wall system, 57% with outrigger system, 58% with outrigger and belt truss system, 30% with shear wall system, 8% with conventional brick infill and 3% reduction with ALC block infill as compared to bare frame. The lateral displacement of all structural systems is going to increase by 50% from one zone to its next higher zone.
2. The maximum storey drift for 40 storeystructure, reduces by 34% in comparison with normal Bare frame structure for a peripheral Shear Wall system, 60% for a Core Wall system, 63% for Outrigger and outrigger with belt truss system, 8% for a conventional brick infill system and 5% for ALC block infill system. Thus it can be concluded that the outrigger with belt truss system structural form is more Sustainable against Displacement due to seismic forces.
3. It is found that the Base Shear for a 40 storey structure, core wall system is reduced by 7%, 6% for outrigger system, 5% for a outrigger with belt truss system, base shear is increased by 50% for conventional brick infill system and 40% for a ALC block infill system as compared to bare frame structural system. The base shear of core wall and outrigger systems was reduced due to presence of opening at the center of core wall.

4. For a 60 storey structure the maximum lateral displacement is reduced by 47% with outrigger and belt truss system, 45% with outrigger system, 40% with core wall system, 22% with shear wall system, 15% with conventional brick infill system and 10% with ALC block infill wall system as compared to bare frame system.

5. For a 60 storey structure the maximum storey drift was reduced by 45% with core wall system, 46% with belt truss andoutriggersystem, 23% with shear wall system, 16% with conventional brick infill system and 9% with ALC block infill system as compared to bare frame structure.

6. Base shear of a 60 storey structure is increased by 30% with conventional brick infill system and 25% with ALC brick infill system, and reduced by 7% reduced with core wall system and 6% with belt truss and outrigger systemas compared bare frame structural system. Base shear is reduced due to presence of opening at the center of core wall.

7. For a 90 storey structure the lateral displacement is reduced by 54% with belt truss and outrigger system, 52% with outrigger system, 48% with core wall system, 15% with shear wall system, 23% with conventional brick system and 20% with ALC block infill system as compared to bare frame structural system.

8. For a 90 storey structure the maximum storey drift was reduced by 47% with belt truss and outrigger system, 2% with core wall system, 24% with conventional brick infill system and 21% with ALC brick infill wall system.

9. The maximum Lateral displacement, Stroey drift and Base shear of a structure is increased by 50% from one seismic zone to its consecutive higher zone(from zone2 to zone3).

10. The lateral displacement of a 60storey structure is 2 times the 40storey structure and 90 storey structure is 5 times and 2.5 times of 40 and 60 storey structure respectively.

11. Storey drift of a 60 storey structure is 1.27 times of 40 storey structure and 90 stroeystructure is 2times of 40 storey and 1.6 times of 60 storey structures respectively.

12. Base shear of a 60 stroey structure is 1.10 times the 40 storey structure and 90 storey structure is 1.6 times the 40 storey structure and 1,4 times the 60 storey structure respectively.

13. As compared to bare frame the infill systems reduces the storey displacement and storey drift. But there is a increase in base shear of the structure. ALC block infill results in little bit more displacement as compared to conventional brick infill systems but more amount of base shear is reduced.

14. Outrigger and outrigger with belt truss has almost same amount of lateral displacement and base shear, belt truss system helps in equal distribution of loads to all columns.

References

[1]   Akshay A. Khanorkar& Mr. S. V. Denge, "Belt Truss as Lateral Load Resisting Structural System for Tall Building: A Review", IJSTE ISSN: 2349–784, **2016**.

[2]   B N Sharath& D Claudiajeyapushpa, "Comparative Seismic Analysis of an Irregular Building with a Shear Wall and Frame Tube system of various sizes", IJECS ISSN: 2319–7242, **2015**

[3]   Momin Mohmedakil M & P.G.Patel, "Seismic assessment of rc frame masonry infill with alcblock", IJERA ISSN:2249–8974, **2012**.

[4]   Muralidhar G.B &Swathi Rani K.S, "Comparison of performance of lateral load resisting systems in multi storey flat slab building", IJRET ISSN: 2319–1163, **2016**.

[5]   P. P. Chandurkar& Dr. P. S. Pajgade,"Seismic Analysis of RCC Building with and Without Shear Wall", IJMER ISSN: 2249–6645, **2013**

[6]   P.M.B. Raj KiranNanduri&B.Suresh, "Optimum Position of Outrigger System for High-Rise Reinforced Concrete Buildings Under Wind And Earthquake Loadings", AJER-ISSN: 2320–0847, **2013**.

[7]   R Shankar Nair, "Belt Trusses & Basements as Virtual Outriggers for Tall structures", Engineering Journal, **1998**

[8]   Srinivas Suresh Kogilgeri& Beryl Shanthapriya, "A Study on Behavior of Outrigger system on High Rise Steel Structure by varying Outrigger Depth", IJRET eISSN: 2319–1163, **2015**.

[9]   Shyam Bhat M & N.A.PremanandShenoy, "Earthquake behaviour of buildings with and without shear walls", JMCE -ISSN: 2278–1684, **2013**.

[10]   Vishal Altware& DrUttam B Kalwane, "Effects of openings in Shear Wall on Seismic Response of Structure", IJERA ISSN: 2248–9622, **2015**

[11]   Vidhya P. Namboothiri, "Seismic Evaluation of RC Building With AAC Block Infill Walls", IJSR ISSN: 2319–7064, **2015**.

[12]   U. L. Salve & R. S. Londhe, "Effect of Curtailed Shear Wall on Storey Drift of High Rise Buildings Subjected To Seismic Loads", IOSR -ISSN: 2278–1684, **2014**.

[13]   Z Bayati, M Mahdikhani& A Rahaei, "Optimized Use of Multi-Outriggers System toStiffen Tall Buildings", Beijing, China, **2008**.

Sandeep Kakde, Pavitha U S, Veena G N, Vinod H C

Implementation of A Semi-Automatic Approach to CAN Protocol Testing for Industry 4.0 Applications

Abstract: In this work, unit testing is done for CAN protocol between two systems in power train i.e., dosing control unit and selective catalytic reduction, by using poly-space and RTRT Tools. In the first phase of the testing, static unit test can be done by using polyspace tool. Polyspace tool can be used to debug the bugs that may occur during run time at early stages. Polyspace provides more tool errors. This drawback can be overcome in S- Function. Functional unit testing of power train system can be done with the help of S Function. For all the auto coded modules, the S-function Unit testing process has been widely accepted and it has been observed significant benefits in the areas of quality as well as the efficiency in the Unit testing stage. The runtime errors can be identified by using polyspace tool and functionality of a block can be verified by using the rational test real time tool.

Keywords: industry 4.0, CAN protocol, s function, polyspace EDA tool, automotive industry

13.1 Introduction

Automotive industry is growing every day as the rates of embedded applications in vehicles are increasing [1]. In order to make sure, each and every functionality in the software is working properly, testing is done before placing the software inside the vehicle. This testing plays a very important role in automotive companies. Since automotive is a safety critical system, driver and passenger safety is important. Testing helps in verifying and validating the code to check it is working as expected or not. If the software is not working as we expected means then tester needs to report bug. The software needs to be bug free. The main problem with existing systems is defined here. If CAN protocol present between the dosing control unit and

Sandeep Kakde, Dept. of Electronics Engineering, Yeshwantrao Chavan College of Engineering, Nagpur, India, e-mail: sandip.kakde@gmail.com
Pavitha U S, Veena G N, Dept. of Electronics and Communication Engineering, M S Ramaiah Institute of Technology, Bangalore, India
Vinod H C, Dept. of Information Science, SJBIT College of Engineering, Bangalore, India

https://doi.org/10.1515/9783110725490-013

selective catalytic reduction unit is not working properly means it will send wrong NOx concentration value to the dosing control unit. This in turn leads to the wrong amount of urea injection to the exhaust chamber. It will result in increased NOx content in the atmosphere.

13.1.1 Introduction to Automotive Electronics

Car hardware is any electrically provided frame used as a part of street vehicles. Car hardware started from the requirement to control motors. Main electronic parts were used to monitor motor capacities and were associated to the motor control units (ECU). As electronic controls were used for more car applications, the acronym ECU went for the more significance of "electronic control unit" and after that a particular ECU's were produced. Now days, ECU's are measured. Writes incorporate motor control modules (ECM) or transmission control modules (TCM). An advanced auto may contain up to hundred ECUS' and business vehicle may contain up to forty. Automotive hardware or car inserted frameworks are conveyed as per distinctive spaces in car field, they can be arrangedinto:
- Powertrainsystems
- Transmission system
- Driverassistance
- Chassis system
- Infotainmentsystems

The main contribution of the work is implementation and testing part.
1. In order to make sure CAN protocol present between the Dosing Control Unit and Selective Catalytic Reduction Unit is working properly, before placing the software inside the vehicle testing is done to detect the bugs.
2. The runtime errors are detected by using polyspace tool and functionality of the blocks present in the specification are tested by using rational test real time tool.
   This paper is organized as follows: Section 1 focuses on introduction part. Section 2 describes the related work. Section 3 mainly focuses on design methodology. Section 4 describes the results and lastly section 5 concludes the paper.

13.2 Previous Work

The automotive sector totally changed the way we lived by opening the ways to private transport and their improvement isn't finished yet. From few years cars have been profoundly changed to take after new bearings and grasp developing patterns [2]. Other areas where automotive industry is growing in anti-brake system, Collision avoiding, air bag, driving modes and reducing emission. Even automotive engineers are working on sensors to get accurate values [3]. Car hardware is any electrically provided frame used as a part of street vehicles. Car hardware started from the requirement to control motors. Main electronic parts were used to monitor motor capacities and were associated to the motor control units.

A standout amongst the most requesting electronic part of a car is electronic control unit which all in entire is depicted as power train framework. Motor controls request one of the most noteworthy ongoing due dates, as the engine is quick and complicated piece of a car [4–5]. Among all the hardware's in any auto the energy consumed by the motor control unit is the most noteworthy; ordinarily a 32-bit processor is used inside ECU. Numerous more engine parameters are effectively observed and controlled progressively. There are around 20 to 50 sensors that measure weight, stream, and speed of the motor, oxygen content and NOx range in addition to different parameters at various focuses inside the motor [6–7]. All these sensor messages are sent to an ECU, which has the rationale circuits to do the genuine controlling. ECU yield is associated with various actuators for the throttle valve, EGR valve, rack [8].

13.3 Sustainable Pollution

Now a day every automotive industry is focused on creating and giving sustainable powertrain advances that takes part in an upgraded life quality of our environment. We lessen immediate and aberrant pollutant outflows from automobile transport, beginning with urban sectors. The 2030 powertrain protocol will be more differentiated than today, containing propelled Integrated Circuit engines, plug-in hybrid electric vehicles, hybrid electric vehicles,and battery electric vehicles. This synthesis of various powertrains will be continuously supplemented by hydrogen fuel cell electric vehicle. Considering this benchmark, the standards of the emission norms are being narrower on exhaust gas emission contents. The

ECU design made for DCU is in accordance to BS-VI norms and the necessary CAN communication of DCU calculated data with other ECUs has also been upgraded since the design complexity has been advanced [2].

13.3 Necessity for Exhaust Gas Post Treatment

Ensured, perfect and beneficial engines will end up being progressively indispensable in present day social requests where we will see bigger measures of transportability on one hand and confined resources of course. Fuel engines for explorer cars have been made to create more power and diminishing out vows meanwhile. As such the engine systems have ended up being eccentric with different subsystems. Because of its unfaltering quality and profitability,the diesel engine is generally preferred for major commercial vehicles as transport, anyway although having starting latency reaching its high torque when used with a turbocharger it has ended up being increasingly conspicuous for major travellers and even transport and high speed vehicles also. The progression of the diesel engines especially the quick infusion and furthermore the essential rail high weight imbuement brought change with respect to power, adequacy, and releases.

Furthermore, exhaust fumes after treatment systems will be made in order to fit in with exhaust level benchmarks like those of fuel engines. To meet the morestringent tests of fuel and diesel engines needs to be unceasing upgraded exhaust aftertreatment design. Nowadays exhaust emission of vehicle engines is usually handled by "DEF".

13.4 Selective Catalytic Reduction System(SCR)

Selective Catalytic Reduction (SCR) is an emission control system as an outcome of combustion that imbues a liquid decreasing agent into debilitates system. An advanced technique that comprises impulse with injector is discharged into the diesel engine's exhaust stream. Reductant agent namely urea which is infused with Diesel Exhaust Fluid (DEF) causes a chemical reaction to occur. This chemical reaction leads to transition of nitrogen oxides compounds into their by-products as water and nitrogen, few percentages of carbon di oxide which is released out of exhaust pipe. SCR advancement is proposed to permit nitrogen oxide reducing reactions underneath permissible dimension. It is called selective as it diminishes percentage of nitrogen oxides. This chemical reaction is known as "reduction" on the grounds that the DEF is the decreasing agent which reacts with nitrogen oxides to transform toxins into nitrogen, water and unassuming measures of carbon dioxide.

13.5 Control Units

CAN is a much acknowledged convention and is utilized in the car businesses from past two decades. It's an essential bus-based convention in the present situation of vehicles. CAN is the control area network, with a high need-based element. The sensor information is traded between the distinctive control units and sensors. Due to substantial number of ECUs and sensors there is a need of legitimate clever convention for message transmission. CAN distinguish the necessary messages and transmits only correct messages between the ECU. With the expansion in number of clients the CAN is additionally altered to CANFD which is known for additional extended bandwidth. LIN is likewise utilized as a piece of correspondence in autos for particularly for non-real time car actuators and sensors. LIN comes with few limitations to bandwidth, so Flex-Ray appeared on account of interest of higherbandwidth.

CAN is appropriate convention for on-going applications and it holds up the bus assertion for messages which hold high priority. It provides instantaneous responses to the data/message that hold more priority. Synchronized communication along with the advent of priority-based data transfer is one of its special features. It lets the information and status share among control units. CAN frame provide development and simple synchronization even if new system requirement is added to the design. Reduced hardware and reduced wiring complexity and weight expand the ease of use of CAN convention in automobile industry.

13.6 Design and Methodology

Here we are testing the CAN frame which is present in the selective catalytic reduction system between dosing control unit and other engine control units. Before passing any information through CAN frame, diagnosis must happens to check for faults and bugs. If no bugs then only process will continues. The sensors present at the end of selective catalytic reduction unit will measure the NOx concentration inside the exhaust chamber and that information is passed to the dosing control unit through the CAN protocol. Before passing through CAN protocol fault diagnosis is performed to detect faults. Once it's confirmed no fault is present then the NOx information will be passed to the dosing control unit ECU. Depending on the received information area will be inserted into the exhaust chamber. This will help to reduce NOx in the atmosphere. Figure 1 shows flow chart of top level units.

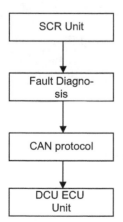

Fig. 1: Implementation of Block Diagram.

13.7 Result and Conclusion

Polyspace is a tool used to find the runtime errors that may occur when vehicle is in running condition. It can be used to find the unreachable theoretical range and shared variable problems. Tool performs the analysis on the required code and produces a report. This report is generated using polyspace bug finder and can be viewed through the polyviewer.

13.8 Polyspace Static Unit Test Report

There are red error, orange error, grey error and green error. If red error occurred means critical error and the execution of further code will be aborted. So in that case we need to correct the code and launch it again. Orange error is less critical error; we can check the code for this and report accordingly. Grey error occurs when function not called or unreachable part of the code is encountered. Green check indicates no error. Fig. 2 shows the polyviewer result window.

13.9 Analysis of Runtime Error

Runtime errors are those which occur during the execution of software. The illegal operations in the software will lead to the run time errors. When the error occurs program will stop executing, so program must be written in such a way that it can handle unexpected errors and instead of terminating unexpectedly, it must continue to execute without any delay. This ability of the software is called as robustness of the code. When system experiences a run time error the execution will take more time. The software will do self-diagnosis and recognizes the run time errors. Fig. 3 shows the possible run time errors. If the run time errors increase means in turn performance of the vehicle will decrease. The following figure show the errors and corresponding comment to the coder.

13.10 Analysis of Unreachable Theoretical Range (UTR) Errors

UTR means Unreachable Theoretical Range, if the variable is taking the range other than the mentioned range in the data dictionary then UTR warning will raise. UTR warnings are present in the second section of the PSTR report. The ranges for the variable are obtained from the view.txt file. The PSTR report will conation the real

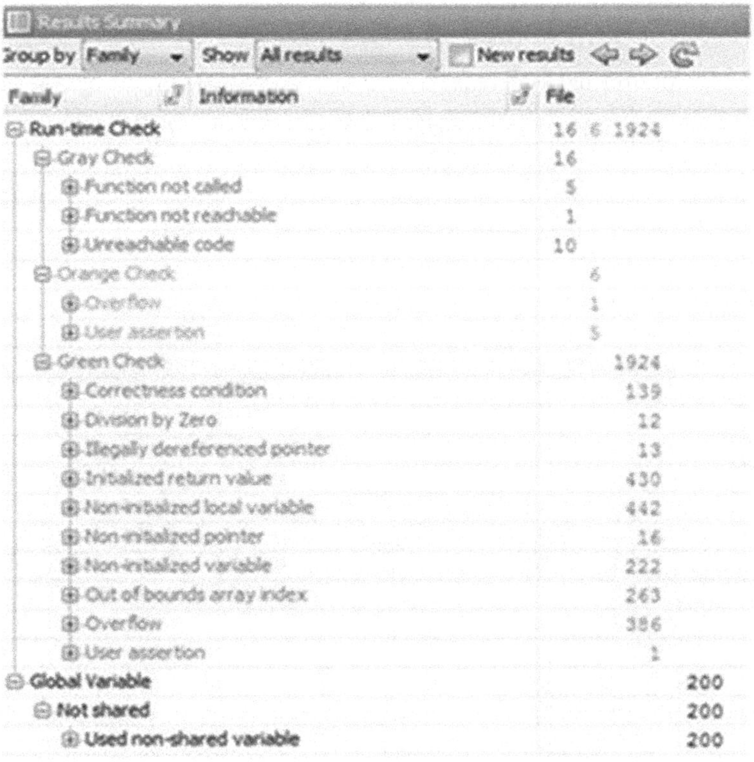

Fig. 2: Polyviewer result window.

range and obtained range of the variable, if the obtained range of the variable is outside the real range then fail error has to be reported. For e.g. clu11_msg_parity_test variable in Fig. 4 has been wrongly limited.

13.10.1 Check Calibration Values

This service in the PTU is used to define all Application Programmable Variable and Application Programming Map variables. In this section one cannot include any external variables mentioned with seeappli into the data dictionary. All variables in this section are assigned with the specification value as per customer requirement. There must be one test present in this service and no function call is required.

RTE warnings (→ Normal project)

Error	Line	Type	Criticality			Error code	Comment
			FAIL	QUAL	PASS		
1	1070	OVFL			X	TOOL_CODE	Unproven: operation [*] on scalar may overflow (on MIN or MAX bounds of INT32) --------- (1070) stack_data_s16_1 = (S16)((stack_data_s32_0 * (1071) stack_data_s16_0) / BIN20); --------- Multiplication result is in the range [535822336, 1071644672] which is within the range of S32[2147483647], so OVFL not occur.
2	1059	OVFL			X	TOOL_CODE	Unproven: operation [conversion from unsigned int16 to int16] on scalar may overflow (result strictly greater than MAX INT16) --------- (1059) stack_data_s16_0 = (S16)stack_data_u16_0; --------- stack_data_u16_0 is updating in range [0..511] which is in S16 range. Hence OVFL not occur.
4	927	OVFL			X	TOOL_CODE	Unproven: operation [conversion from unsigned int32 to unsigned int16] on scalar may overflow (result strictly greater than MAX UINT16) --------- (927) stack_data_u16_0 = (U16)GET_BTFLD_V1 (928) (&(buffer_clu[0]), (929) (U8)1, (930) (U8)0, (931) (U8)9); --------- GET_BTFLD_V1 function give o/p from [0..511] which is in U16 range. Hence OVFL not occur.

Fig. 3: Polyspace run time errors and comments.

UTR warnings (→ Normal project)

Error	Line	Type	Criticality			Error code	Comment
			FAIL	QUAL	PASS		
1	-	UTR			X	PASS_CODE	clu11_msg_parity_test DD : [0 .. 15] - Real : [0 .. 9] Implemented as per requirement.

Fig. 4: Polyspace unreachable theoretical range error and comment.

13.10.2 Principle

The specification is created by using spec tool; Check that value assigned to the variable is same as that of one defined into the specification by the customer. Check spec DD command is used to verify Break Point-X input for map and Break Point-Y input for map values.

13.10.3 Check for Initializations

This service is dedicated to verify the initialization function. This service main goal is to ensure, output interface variables must be initialized with the values provided into the specification given by the customer. This service also checks that variables that are related with memory effect are also initialized. It is also necessary to check that nvv variable is not initialized in the initialization section. Usually there must be only one test present in the initialization section and one call to the initialization function.

13.10.4 Check for MainFunctions

Objective that must be kept in mind while building RTRT test is that at the end of the execution, full coverage between the specification and the code must be controlled. The main function is called inside each test case other than initialization test case.

13.11 Coverage Result

This is the last section in the Atoll test report. This section gives details of source code that has been uncovered by the tester on mistake or by purposefully. It is a

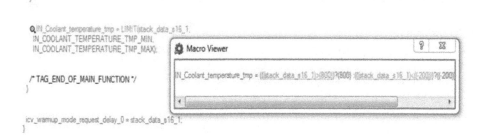

Fig. 5: Uncovered part of the code.

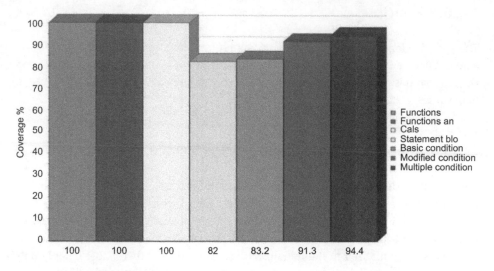

Fig. 6: Graphical representation of little code coverage.

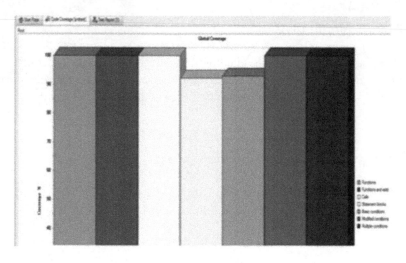

Fig. 7: Graphical representation of full code coverage.

responsibility of the tester to justify all the uncovered code comments elobratively. If the report is prepared means, tester can deliver the report as shown in the Fig. 5.

Coder needs to do any changes that time they will go through the reported ATR. At the end of the test construction, there may be some parts of the code uncovered while testing which can be reported as follows. Figs. 7 shows full code coverage histograms.

13.12 Conclusion and Future Work

In automotive industry, competition is increasing day by day as the electronics in the vehicle is increasing. So it is important to verify the software and validate the system before it send to the production. Since automotive is the safety critical system, we must do the testing. The runtime errors can be identified by using polyspace tool and functionality of a block can be verified by using the rational test real time tool. In rational test real time tool, we write test case to identify the functional problems with the code. If the code is not correct then tester need to report so that coder will understand this and correct in the future version of the code.

Developer and tester will lack in time to deliver the products on time to the client. Time is the main constraint in the present world of automation due to the immense competition from the other organizations. So coder needs to write the code as soon as possible and the tester can test the code after coder writes the code. It will take lot of time and the organization may fail to deliver the products on time. Making automation in scripts can help coder and tester to do the task fast by taking less time. So that delivery of the product can be done as per the required time. Also it will take less cost.

References

[1] Manoj Rohit Vemparala, Shikhara Yerabati, Gerardine Immaculate Mary, "Performance Alysis of Controller Area Network based Safety System in an Electric Vehicle", IEEE International Conference On Recent Trends In Electronics Information Communication Technology, May 20–21. (2016).

[2] Khan O, Mueller F. "Recent Advances in In-vehicle embedded systems". InWSA 2017; 21th International ITG Workshop on Smart Antennas; (2017) pp.1–4.

[3] Vrushali Shinde, Prof.S.P.Karmore,"Trends in Automotive Communication Systems" IEEE Journals,(2005).

[4] Y. Kanehagsi, D. Umeda, A. Hayashi, K. Kimura, H. Kasahara. "The design of safe automotive electronic systems Symposium on Industrial Embedded Systems". (2006).

[5] Patil M. "Embedded Software Product Verification & Validation Using Virtual Reality" IEEE International 2015 Aug 27,(2015).

[6] Socci V. "Implementing a model-based design and test workflow In Systems Engineering (ISSE)", 2015 IEEE International Symposium, Sep 28 2015,pp. 130–134.

[7] Boyang Du and Luca Sterpone Dipartimento di Automatica Informatica Politecnico di Torino Italy, "An FPGA-based Testing Platform for the Validation of Automotive Power train ECU" IFIP/IEEE International Conference on Very Large Scale Integration. (2016).

[8] Pavankumar Naik, Arun kumbi, Nagaraj Telkar, Kiran Kotin, Kirthishree C Katti, "An Automotive Diagnostics, Fuel Efficiency and Emission Monitoring System Using CAN", (2017) IEEE.

[9]   Du, B. & Sterpone, L. (2016), "An fpga-based testing platform for the validation of automotive powertrain ecu, in Very Large Scale Integration (VLSI-SoC)", IFIP/IEEE International Conference on', IEEE, pp. 1–7,(2016).

[10]   Yu, J. & Wilamowski, B. M. "Recent advances in in-vehicle embedded systems", in IECON 2011–37th Annual Conference on IEEE Industrial Electronics Society, IEEE, pp. 4623–4625, (2011).

[11]   Navet, N., Song, Y, Simonot-Lion, F. & Wilwert, C. "Trends in automotive communication systems", Proceedings of the IEEE 93(6),pp.1204–1223,(2005).

Dwarakanath G V, Dr P Ganesh, Sinchana S R, Ashwini C R

Smart Cradle System for Industry 4.0

Abstract: This chapter is aimed at developing an interactive "SMART CRADLE SYS-TEM". People are getting busy with their work; no one has time to look away from earning money. Parents don't have enough to look after their babies. The Mother has to take care of their baby and also work for their office concurrently. A cradle system is going to help the parents to rest, even if both of them go to work or woman is housewife, taking care of a baby is stressful. When people are in stress it creates a bad environment around the baby, so being stress free is most essential. A smart cradle system will swing on its own even if no one is swinging it, when a baby cries and it also sends an alert message to parents so they can be notified. Sometimes, we don't notice if the baby has urinated, this has to be cleaned so that the baby doesn't get infected. Checking every time is not required, our smart cradle system will send an alert message, if the cradle is wet. Proposed chapter is going to the parents to take care of their baby by taking rest and have a stress free environment.

Keywords: gsm module, smart cradle, sound sensor, wet sensor

14.1 Introduction

The chapter is titled "SMART CRADLE SYSTEM" is an IoT based project, Internet of Things (IoT) is an emerging technology which is used to save our time and reduce human intervention with machine. With the help of IoT, we are trying to save the time of the parents and security of the baby, we have built a smart cradle system which will not only update parents about their babies' behavior but also give a stress free environment. Keeping the baby safe and secure is our main intention.

In today's world, most of the families are nuclear family, which consist of parents and their children. Generations back they were grandparents, uncle and aunt, etc. Because of this new form of family i.e. a nuclear family looking after the new born baby, it is the duty of a mother. In this modern era, people are ready to adopt modern technologies. So to ensure babies health, comfort, secureness, a SMART CRADLE SYSTEM is proposed which is an automated cradle system connected to the parents mobile to monitor a baby. Sound of a baby is fed as input to sound sensor which is attached to cradle. It considers only the sound of baby, it takes the decibels, if the sound is more than the threshold value, the swing starts swinging the cradle, if the baby doesn't stop

Dwarakanath G V, Dr P Ganesh, Sinchana S R, Ashwini C R, Department of Master of Computer Applications, BMSIT&M, Bangalore, 560064, India, e-mail: ambikaganesh@gmail.com

https://doi.org/10.1515/9783110725490-014

crying, an alert message is sent to the parents to notify them. If the baby has urinated, the cradle system has a wet sensor which detect the wetness in the cradle, it sends an alert message to parents. This cradle system will help the parents by looking after the baby and create a stress free environment.

Cradle system will give working parents to rest as the smart cradle take care of the baby in the absence of parents. The cradle system will alert the wetness in the cradle to parents to change or clean baby when it has done pee. This system will automatically start to swing if the baby is crying. Parents being stress free will create great atmosphere around the baby.

The operation sequence of smart cradle system is given a follows: If the baby's sound is detected or baby is making noise then sound sensor triggers cradle to swing. Also an alert message is sent to parents through blynk app. And the second functionality is, if the baby had done pee and wetted the matrices of the cradle then parent will be alerted to change it.

14.1.1 Problem Statement

An infant while growing up all it wants is love and care of its parents, it can never replace by the wealth etc. Parents has to take care of them and also do their work concurrently, this can be overload and create stress environment. A mother has to do all the work, even if she is a housewife it will be stressful to do all the work by herself. The mother has to get enough sleep to be healthy, in order for the baby to be health the mother has to be healthy because she feeds the baby. Our smart cradle system is here for her help.

Some infant monitoring system was developed to notify the parents during alarm conditions. But this cradle system is developed to notify the one or the other activity of the baby. It is implemented to help the parents who were busy in office work. In our proposed cradle system can sense the wet conditions of the baby's bed and it also detects the crying sound. Initially, data is collected from the sensors and it is monitored with subjected values. Finally, alert messages sent to parent mobile phones during abnormal conditions. The measured various parameters of sensors were displayed on the mobile phone. The cradle will start swinging by using the servo motor when the crying sound gets detected by the sensor [11].

14.1.2 Objective of the Project

- We are developing a smart cradle system which will automatically swing the cradle and check the bed-wet condition.
- We can operate it easily and this an reduce the work of human.
- To build a cradle system for a baby which is safe and comfortable, it will swing when a baby cries and also detect if the baby urinates.

– To build a cradle which is cost efficient and more flexible.
– User friendly.

14.2 Literature Survey

From the normal cradle to the automatic swing cradle, there are many types of the cradle. The use of baby cradle came in when the security of the baby was at risk, it would be more cozy for a baby to sleep in a cradle or in confinement which is specially made for them with extra bedding such as blankets or quilts. It was further modified in such a way that they are portable. Looking after a baby is a difficult task in the initial stages. Baby monitoring system will help parents to try and understand the uneasiness experienced by the baby. Mainly, when the parents are not nearby, the swinging motion helps baby relax and take a short nap. This was why some automatic swinging cradle was developed.

As we all know, there are baby day care centers and nannies who were getting paid to take care of baby or sooth the baby. We should be aware of the fraud detected in these kinds of centers. Usually the cradle systems have one or two features which is also make the baby uncomfortable by disturbing the baby, if the baby urinate, but our smart cradle system help the baby to have peaceful sleep and also after the hygiene of the baby.

There are many home-care which are designed for adults which will monitor their health stats, emergency signals are sent if there is any discomfort for any one present there. But their functionality is different from taking care of a baby. An infant can't give feedback or tell it out that they are having problems. The system used in taking care of the adults can't be implemented for taking care of the infant. Majority of the parents are not comfortable to leave their kids in nursing home. In this generations, new viruses or diseases are emerging so leaving the kids away from home is not safe. Many research papers and patents for the well-being of the infants are being done to find solution to look after them.

14.2.1 Survey of Papers

First paper is [1]. "**Smart baby cradle**" is an IoT project to build a cradle system which was published in 2018. This paper has features like PCB for sensing wet conditions, cradle swing which will move side to side, this will help the baby to sleep, this is done by geared motor. They have also installed Arduino camera for watching the baby on live. They have used other sensors like PIR sensor to detect the light level inside the cradle, etc. it helps the parents to monitor the baby, checks the temperature of the infant, wetness will be detected by sensor.

Second paper is [2]. **"Development of an Intelligent Cradle for Home and Hospital Use"** published in the year 2015. This system is designed to monitor baby movement, bed-wet condition of baby and body temperature. Infant's body movement is detected by using PIR sensor. The baby's body temperature is recorded and caution is sent to parents if body temperature goes above the given threshold temperature. In addition to this a wet sensor is also incorporated to notify the folks or the attendants for change the of the bed. It is going to keep the surrounding of the baby clean and hygienic. This doesn't have the main module swinging of cradle.

Third paper is [3] **"An Automatic Monitoring and Swing the Baby Cradle for Infant Care"** published in 2015. In this system, they have installed reading facial expression system which determines the baby is safe. In this system there is cry analyzing which detects if the baby is crying, the swing will start swinging on its own till the baby stops crying. There is alarm system which will ring for two purpose, that is, if the baby doesn't stop crying even after the cradle system swings and if the baby urinates and cradle gets wet.

Fourth paper is [4]. **"IOT Based Smart Cradle System for Baby Monitoring"** was published in 2019. This research paper, the author has added features like PIR sensor to detect any motion which is used for security purposes. They have also used two temperature sensor, those are LM35 and DHT11. The DHT11 sensor will check the whole room's temperature and the LM35 sensor will check the baby's temperature. If there is sound more than a certain amount, there is sound sensor which will detect it and it will automatically swings the cradle. If the baby urinates in the cradle the wet sensor will detect it.

Fifth paper is [5]. **"A Continuous Infant Monitoring System Using Iot"** The research paper, the author has added features like PIR sensor, sound sensor, temperature sensor and humidity sensor. This system will monitor the temperature of the baby, detects if there is any insects and any flies around the baby, if the diaper of the baby gets too wet, it will send the notification to change it, through Wi-Fi module. We can also check the condition of the baby through this system using Blynk application.

14.2.2 The Existing System

The cradle is used to make the baby to sleep, the baby is put into this cradle when the baby needs to rest and the parents can work with their daily routine. But these conventional cradle have to be pushed by someone to make the baby sleep. These conventional cradle systems are not automatic; this need manpower for them to work. These conventional cradle are used in non-developed cities and villages because of their low prices. The problem with these conventional cradle is that it is not always safe and some might not be comfortable to the baby. So we need an automatic cradle system to look after the baby when they are put into the cradle to rest.

There are already existing products like "Smart baby cradle", "Intelligent baby cradle" these are products which are conventional model wherein they use mechanism to make the baby sleep. These cradle mechanism moves from East to West which will lead to Shaken Baby Syndrome (SBS). This starts to shear and tear motion between those tissues of white and grey matter of the brain which serves as the major cause of the damage of the IQ and cognitive function.

14.2.3 The Proposed System

We are proposing this project to break the challenges faced by the above existing systems and develop a smart cradle system to assist the baby keeping them safe and comfortable. Our system is designed in a such way the cradle will move in the side-to -side, by using this mechanism the baby is prevented by SBS. The main purpose of this project is to help the parents of the babies to look after their baby even if they are working. The gap between the parents and the baby is no more. The various sensors attached to this cradle system will keep monitoring the activities of the baby, that the cradle system will swing the cradle automatically that there is no need for any manpower to swing. If problem is caused then parents are intimated by sending the notification to their mobile via blynk app, the proposed system will help the parents by decreasing their stress of looking after the baby, many hurdles are being solved by our system. The baby will be monitored all the time when it is in the cradle, it is healthier and safe. The babies will have a comfortable safe environment to sleep.

14.3 Software-Requirements-Specification

The requirements of the end users that are demanded specifically as basic facilities that a system can offer is called as software requirements specifications. A functional requirement defines a system or its component. Functional requirements of our project explains the functionality that must be provided by each module of our project.

14.3.1 Function of a Servo motor

These Servo Motors are small devices which has an output shaft. The shaft in these devices can be placed in a position to an particular angular position which sends a coded signal to the servo. The servo maintains the shaft's angular position as long as the coded signal is on the input line. The shaft changes when the codes signal are changed.

Here, in our project function of servo motor is used for swing the cradle, when baby starts crying the cradle will automatically starts swinging. These motors are electric devices which are used for the objects to rotate precisely, the rotational angle lies between from 0^0 to 180^0, we are using this method to swing the cradle. This will move cradle side-to side.

14.3.2 Function of a Wet Sensing Sensor

Wet sensor is used to detect if there is any kind of liquid falling on the sensor, this is caught by the Arduino board from which we can perform required action. We can use this method is varying fields. Wet sensor here it is utilized for distinguishing the wetness of the diaper, when the diaper is wet, the notice is sending to the guardian's portable. The support is structured so that to make the child more solace or comfort.

If the baby urinates, the sensor will detect it. The sensor functionality in this project is to check if the baby's mattress is wet. When the wet is detect the parents will be notified by a message through blynk app. This will provide a hygienic environment for the baby.

14.3.3 Function of a Sound Sensor

Sound sensor is portrayed as a module that recognizes sound waves through its power and changing over it to electric signs. A Sound Sensor is an essential framework that recognizes sound. It is basically situated a Microphone with some managing circuit. Using a Sound Sensor, you may measure the power of sound from different sources like bangs, acclaims, riotous voices, etc

In this project the primary reason for the cry discovery unit is to identify the voice of the infant when an infant is crying. At whatever point the crying sound gets identified the support gets swings by utilizing a servo engine.

Sound sensor is used for taking the baby's sound as an input, this is done by taking the range of decibel the sound of the baby, if the range is rectified by the sensor then a signal is sent to the motor which will automatically swing the cradle.

14.3.4 Function of Arduino Uno

Arduino is software and hardware based open-source platform which is easy to use. These boards can read the inputs and make them as required outputs. The board function is to send a set of instructions to microcontroller. We use the Arduino Programming language, IDE which is easy to program. Arduino Uno is microcontroller board. ATmega328 which has 14 and 6 digital and analog pins respectively.

This board will require an external power to operate, in this project we are taking inputs from wet sensor, sound sensor. Used to connect the sensors of the cradle system, takes the signal from them and controls the whole system.

14.3.5 Function of Nodemcu ESP8266 module

Nodemcu is a development board specifically target on IoT based application. The firmware which runs on ESP8266 Wi-fi SoC from Espressif System, the hardware which based on the module ESP-12. It was later supported ESP32, the 32 bit MCU that was added. Nodemcu is firmware which was open source prototyping board design are available.

Node and MCU is from where NodeMCU which speaks about the firmware than associated development kit. It is used to send the data obtained from these sensors from our smart cradle system through Blynk app Adriano.

14.3.6 Function of a Blynk App

Blynk is a new purposed platform which will be help to build a interfaces for monitoring and controlling the hardware project from devices such as iOS and android. This will help create great interfaces for the project which uses widgets we use. We can create dashboards and these buttons, graphs, sliders and other widgets into the screen. These widgets will helpful, the pins that are connected can be turned on and off or the data can be displayed from these sensors. Blynk server are responsible to communicate between the hardware and smartphones. We can use the blynk cloud or we can run the private Blynk server locally.

Personally, I'm using it to send the notification to the user mobile. When the baby cries the cradle starts swinging automatically till the baby stops to cry. This is helped by the sound sensor to detect and notification will be sent to parents or a caretaker's mobile through blynk app by sending a message that "baby is crying". Even when any wetness is sensed by the wet sensor then the message will be intimated to the parents or caretakers mobile through blynk app by sending message that "baby bed got wet".

14.4 Working Methodology

System Design

Design is the first step into the development phase for any engineered product or system. Design is a creative process. A good design is the key to effective system. The term "design"is defined as "the process of applying various techniques and principles for the purpose of defining a process or a system in sufficient detail to permit its physical realization". It may be defined as a process of applying various techniques and principles for the purpose of defining a device, a process or a system in sufficient detail to permit its physical realization. The system design develops the architectural detail required to build a system or product.

The design phase is a transition from a user oriented document to a document to the programmers or database personnel. System design goes through two phases of development: Logical and Physical Design.

Overview of the system

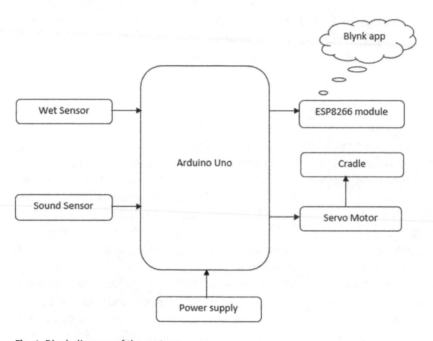

Fig. 1: Block diagram of the system.

Arduino Uno is the heart of the system. Arduino Uno controls the whole system. When the baby is awake from the sleep, initially the cradle swings automatically with the help of the servo motor respectively. This may help the baby to stop crying and go back to sleep once again. But in case the baby does not stop crying, the situation is notified to the parent. Also the system checks the wetness of the bed. If there is a change then wet sensor corresponding actions will be carried out. If the wetness is detected, the blynk notifies the parent.

The above block diagram depicts the complete working of the smart cradle system. When the child is made to sleep on the cradle various sensors like sound detection sensor, wet sensor and servo motor are implemented to monitor the various actions of the child figure 1.

The above diagram shows the automatic working of the cradle. When the sound of the baby cry is detected the system checks whether the sound level is above the threshold value set initially. If the sound is above the mentioned threshold value a signal is sent to the servo motor for the automatic swing and if the baby is crying a notification is sent to the blynk application that the baby is crying. When the baby is sleeping the sound will be less than the threshold value and if the wet sensor sense any kind of wetness in the bed then a notification is sent to the blynk application.

Arduino Uno

Fig. 2: Arduino uno.

Arduino Uno is microcontroller board. ATmega328 which has 14 and 6 digital and analog pins respectively., it has an USB connection, has external power supply jack, an ISCP and also a reset button. Power supply has to give externally. Here, this board will require an external power to operate, in this project we are taking inputs from wet sensor, sound sensor. Used to connect the sensors of the cradle system, takes the signal from them and controls the whole system.

Servo Motor

Fig. 3: Servo motor.

These Servo Motors are small devices which has an output shaft. The shaft in these devices can be placed in a position to an particular angular position which sends a coded signal to the servo. The servo maintains the shaft's angular position as long as the coded signal is on the input line. These motors are electric devices which are used for the objects to rotate precisely, the rotational angle lies between from 00 to 1800, we are using this method to swing the cradle. This will move cradle side-to side.

Wet Sensor

Fig. 4: Wet sensor.

The wet sensor will detect if there is any liquid is on the sensor. The sensor functionality in this project is to check if the baby's mattress is wet. When the wet is detect the parents will be notified by a message through blynk app. This will provide a hygienic environment for the baby.

Sound Sensor

Fig. 5: Sound sensor.

Sound sensor is used for taking the baby's sound as an input, this is done by taking the range of decibel the sound of the baby, if the range is rectified by the sensor then a signal is sent to the motor which will automatically swing the cradle.

Nodemcu ESP8266 Module

Fig. 6: Nodemcu ESP8266 module.

Nodemcu is a development board specifically target on IoT based application. The firmware which runs on ESP8266 Wi-fi SoC from Espressif System, the hardware which based on the module ESP-12. It was later supported ESP32, the 32 bit MCU that was added. Nodemcu is firmware which was open source prototyping board design are available. It is used to send the data obtained from these sensors from our smart cradle system through Blynk app Adriano.

Automatic Cradle Swing and Sound Detection

When the baby starts crying, the cradle will start swinging automatically, this is done by detecting the sound by sound sensor and it will send an alert notification to the parents.

Flow Diagram

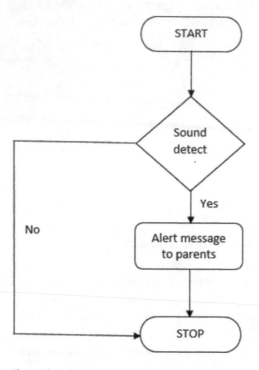

Fig. 7: Flow diagram.

Wetness Detection

Wet sensor is used to detect baby's wetness condition. So this sensor keeps on monitoring for baby's mattress wet condition. Mattress should either be wet or dry. If the wetness is sensed by the wet sensor, then the parents will be alerted with a message to change it. By this baby cannot have any infection or allergy and it will be in a very hygienic environment.

Condition Flow diagram

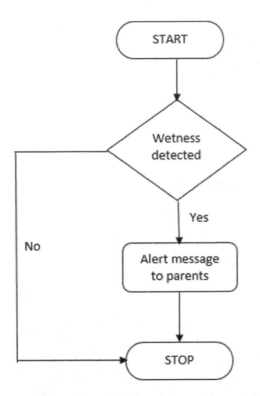

Fig. 8: Condition Flow diagram.

Fig. 9: Sensor Connectivity.

Fig. 10: Sensor and Board Connectivity.

14.5 Conclusion

The growth of technology has been increased rapidly and its contribution to the society can be done in various way. For every parent their kid is most precious. As time flies, the days spent with their little ones seem long. So to ease and nourish their baby, parents spend excess amount of money. As the parents are busy with their works, finding the good and genuine child care is a big deal for them. Our proposed project Smart Cradle System ensures safety of the baby inside the cradle. Keeping cost and safety of baby in mind, cradle comes with more features, less expensive and secure. Another factor with which parents will always be worried about is baby's health. So the cradle system is developed to serve the baby's health factor as well. Smart cradle system lets the parents to do their job besides taking care of the baby at a time. When the baby cries, cradle will start to swing automatically and parents are intimated by sending message and if any kind of wetness is detected then alert will also be sent to the parents. This smart cradle is affordable, reliable, can be maintained easily, safe to use than other automatic electronic cradle.

14.6 Future Scope and Enhancement

We can add features in the future to the cradle system to make work more efficiently and user friendly.

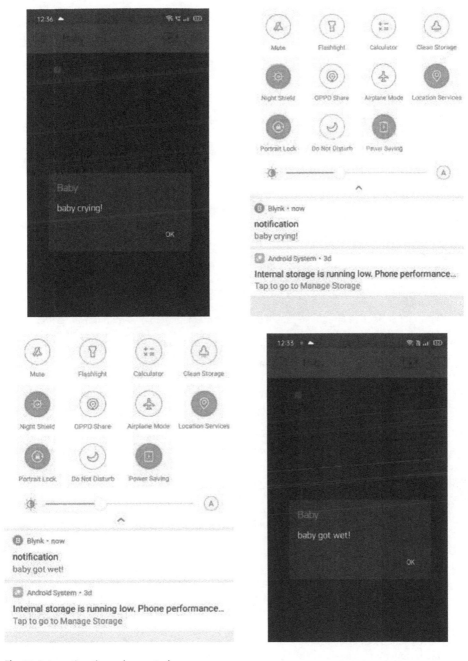

Fig. 11: Interaction through smart phone.

Features like toys which can rotate automatically, music to calm down the baby and live camera via 3G for parents to monitor their baby.

– Camera for live video footage.
– Features like PIR sensor for motion detection, to check if the baby is present in cradle all the time.
– Temperature sensor for measuring babies body temperature and also check the whole room temperature.

The technology has been emerging daily to ease our daily routine life of parents along with their baby's security.

References

Books

[1]   "Smart Baby Cradle an IOT based Cradle Management System." By Prof. A. R. Patil,, International Conference on Smart City and Emerging Technology (ICSCET), 2018.

[2]   "Development of an Intelligent Cradle for Home andHospital Use", Aquib Nawaz, National Inst. of Technology, 2015.

[3]   "An Automatic Monitoring and Swing the Baby Cradle for Infant Care", Rachna Palaskar, Shweta Pandey, Ashwini Telang, Akshada Wagh, Ramesh R. Kagalkar, Int. J. of Advanced Research in Computer and Commun. Eng., Dec 2015.

[4]   "IOT Based Smart Cradle System for Baby Monitoring." Harshad Suresh Gare, Bhushan Kiran Shahne, Kavita Suresh Jori, Sweety G Jachak., International Research Journal of Engineering and Technology (IRJET), 2019.

[5]   "A Continuous Infant Monitoring System Using Iot ", S. Srithar, C. Ravindran, S. Prasad, K. Praveen Kumar, K. Santhosh, International Journal of Future Generation Communication and Networking, (2020).

[6]   "Telehealth mobile system ", J.E. Garcia, R.A. Torres, IEEE Conference publication on Pan American Health Care Exchanges, May 4, 2013.

[7]   "An Embedded, GSM based, Multi parameter, Real-time Patient Monitoring System and Control", Nitin P. Jain, Preeti N. Jain, and Trupti P. Agarkar, IEEE Conference publication in World Congress on Information and Communication Technologies, Nov 2, 2013.

[8]   "Body Temperature and Electrocardiogram Monitoring Using SMS-Based Telemedicine System", Ashraf A Tahat, IEEE international conference on Wireless pervasive computing (ISWPC), 13 Feb 2009. [4]. "Design of a Home Care Instrument Based on Embedded System", Jia-Ren Chang Chien, IEEE international conference on industrial technology(ICIT), 24 April 2008.

[9]   "Low Cost Infant Monitoring and Communication System", Elham Saadatian, Shruti Priyalyer, Chen Lihui, Owen Noel Newton Fernando, Nii Hideaki, Adrian David Cheok, AjithPerakum Madurapperuma, Gopalakrishnakone Ponnampalam, and Zubair Amin, IEEE international conference publication,Science and Engineering Research, 5–6 Dec. 2011.

[10] "Portable Wireless Biomedical Temperature Monitoring System", Baker Mohammad, Hazem Elgabra, Reem Ashour, and Hani Saleh, IEEE international conference publication on innovations in information technology (IIT), 19 March 2013.

[11]   "Development of Optimal Photosensors Based Heart Pulse Detector", N. M. Z. Hashim, International Journal of Engineering and Technology (IJET), Aug-Sep 2013.

Online Reference

[1]   www.pantechsolutions.net/automatic-baby-bed-using-arduino(Accessed 10/06/2021)
[2]   http://myarduinoprojectsvivektr.blogspot.com/ 2016/10/automatic-e-baby-cradle-swing-based-on.html (Accessed 10/06/2021)

Editors Details

Dr. Niranjanamurthy M, Assistant Professor, Department of Computer Applications, M S Ramaiah Institute of Technology (Affiliated to Visvesvaraya Technological University, Karnataka), Bangalore, Karnataka. He did Ph.D. Computer Science at JJTU, Rajasthan (2016), MPhil-Computer Science at VMU, Salem (2009), Masters in Computer Applications at Visvesvaraiah Technological University, Belgaum, Karnataka (2007). BCA from Kuvempu University 2004 with University 5th Rank. He has 12* years of teaching experience and 2 years of industry experience as a Software Engineer. Published 8 books in Scholars Press Germany and One book under process in CRC press. Series Editor in CRC Press and Wiley SP. Published 70+ research papers in various National / International Conferences and Reputed International Journals. Filed 22 Patents in that 4 granted. Currently he is guiding four Ph.D. research scholars in the area of Data Science, Edge Computing, ML, and Networking. He is working as a reviewer in 22 International Journals. Two times got Best Research journal reviewer award. Got Young Researcher award- Computer Science Engineering – 2018. Worked as National/ International Ph.D. examiner. He was conducted various National Level workshops and Delivered Lecture. Conducted National and International Conferences. Member of Various Societies IAENG, INSC, IEEE etc. areas of interest are Data Science, ML, Augmented and Virtual Reality, Edge Computing, E-Commerce and M-Commerce related to Industry Internal tool enhancement, Software Testing, Software Engineering, Web Services, Web-Technologies, Cloud Computing, Big data analytics, Networking.

Dr. Sheng-Lung Peng, Professor, National Taipei University of Business (Professor, Department of Creative Technologies and Product Design) Taipei Campus: No. 321, Sec 1, Jinan Rd, Zhongzeng Dist, Taipei City 10051, Taiwan. Education: Doctor of Philosophy in Computer Science, National Tsing Hua University, Taiwan, 1999. Thesis: A study of graph searching on special graphs Advisors: Chuan Yi Tang / Ming-Tat Ko Master of Science in Computer Science and Information Engineering, National Chung Cheng University, Taiwan, 1992. Thesis: Efficient algorithms for domatic partition problem on interval graphs and their extensions Advisor: Mao-Hsiang Chang Bachelor of Science in Mathematics, National Tsing Hua University, Taiwan, 1988. Published more than 80 research papers and 50+ Books. Areas of interest are Data Science, AI, Edge Computing, Software Testing, Business and Economics, Web Services, Web-Technologies, Big data analytics, Networking.

Dr. Naresh E. is a PhD in CSE and M.Tech degree holder and is serving as an Assistant Professor in ISE Department of MSRIT. Bangalore, India. He is interested in subjects related to Software Quality Engineering, Machine Learning, and Data Science. Software Engineering. AREA OF INTEREST: Operating Systems, Data Mining, Software Quality Engineering Software metrics and measurements, Member of ISTE – LM67925, IAE – 139036, ACM – 8192769, ICSES – 465, ACEEE – 1725. Published 30+ research papers in various National / International Conferences and Reputed International Journals. Filed many Patents.

Dr. Jayasimha S R, Assistant Profesoor, Department of Computer Applications, RV College of Engineering (Visvesvaraya Technological University), Bangalore. INDIA. Having 10 Years of teaching experience. Publications: International Journal: 04, International Conference: 05, National Conference: 14, Book Publications: 01. Areas of interest are Cloud Computing, Data Science, ML, Edge Computing, Software Engineering, Web Services, Web-Technologies, Big data analytics, networking.

Dr. Valentina E. Balas is currently Full Professor in the Department of Automatics and Applied Software at the Faculty of Engineering, "Aurel Vlaicu" University of Arad, Romania. She holds a Ph.D. in Applied Electronics and Telecommunications from Polytechnic University of Timisoara. Dr. Balas is author of

https://doi.org/10.1515/9783110725490-015

more than 270 research papers in refereed journals and International Conferences. Her research interests are in Intelligent Systems, Fuzzy Control, Soft Computing, Smart Sensors, Information Fusion, Modeling and Simulation. She is the Editor-in Chief to International Journal of Advanced Intelligence Paradigms (IJAIP) and to International Journal of Computational Systems Engineering (IJCSysE), member in Editorial Board member of several national and international journals and is evaluator expert for national, international projects and PhD Thesis. Dr. Balas is the director of Intelligent Systems Research Centre in Aurel Vlaicu University of Arad and Director of the Department of International Relations, Programs and Projects in the same university. She served as General Chair of the International Workshop Soft Computing and Applications (SOFA) in eight editions 2005–2018 held in Romania and Hungary. Dr. Balas participated in many international conferences as Organizer, Honorary Chair, Session Chair and member in Steering, Advisory or International Program Committees. She is a member of EUSFLAT, SIAM and a Senior Member IEEE, member in TC – Fuzzy Systems (IEEE CIS), member in TC – Emergent Technologies (IEEE CIS), member in TC – Soft Computing (IEEE SMCS). Dr. Balas was past Vice-president (Awards) of IFSA International Fuzzy Systems Association Council (2013–2015) and is a Joint Secretary of the Governing Council of Forum for Interdisciplinary Mathematics (FIM).

Index

algorithm 22
algorithms 42
automation 15, 66

blockchain 51, 55, 57–59, 89

challenges 121
characteristic 108
code 193
computer 132
computing 55, 105
concentration 182–183, 186

Deep Learning 1, 4
design 169, 173, 177
designing 192
digitization 13
distributed 67, 70, 75, 87
Drones 120

edge 190–191
effective 155–156

Intelligent 218
IoT 119, 120–123, 125–128

LCD 113

manufacturing 43
methodology 132–140, 147, 149

organizations 53

parameters 116, 191, 196
products 155–156, 167
progressive 112
protocol 203–205, 207

Quality 180

rainwater 169

sensor 205–206, 215–218, 220–226, 230, 220, 224
smartphone 66
software 203–204, 208, 213, 219

trends 26

utilization 133–134, 137, 169, 177

https://doi.org/10.1515/9783110725490-016

De Gruyter Series
Smart Computing Applications

ISSN 2700-6239
e-ISSN 2700-6247

Cloud Security
Techniques and Applications
Edited by Sirisha Potluri, Katta Subba Rao, Sachi Nandan Mohanty, 2021
ISBN 978-3-11-073750-9, e-ISBN (PDF) 978-3-11-073257-3, e-ISBN (EPUB) 978-3-11-073270-2

Knowledge Management and Web 3.0
Next Generation Business Models
Sandeep Kautish, Deepmala Singh, ZdzIslaw Polkowski, Alka Mayura, Mary Jeyanthi, 2021
ISBN 978-3-11-072264-2, e-ISBN (PDF) 978-3-11-072278-9, e-ISBN (EPUB) 978-3-11-072293-2

Knowledge Engineering for Modern Information Systems
Methods, Models and Tools
Anand Sharma, Sandeep Kautish, Prateek Agrawal, Vishu Madaan, Charu Gupta,
Saurav Nanda, 2021
ISBN 978-3-11-071316-9, e-ISBN (PDF) 978-3-11-071363-3, e-ISBN (EPUB) 978-3-11-071369-5